Who Cares About Wildlife?

Michael J. Manfredo

Who Cares About Wildlife?

Social Science Concepts for Exploring
Human-Wildlife Relationships
and Conservation Issues

 Springer

Michael J. Manfredo
Colorado State University
Fort Collins, CO 80523
USA
michael.manfredo@colostate.edu

ISBN: 978-0-387-77038-3   e-ISBN: 978-0-387-77040-6
DOI: 10.1007/978-0-387-77040-6

Library of Congress Control Number: 2008932523

Cover illustration: Photo courtesy of www.brilliantexpressions.com

Printed on acid-free paper

springer.com

My orderly burst into my office to tell me that a prisoner had escaped from the brig. The prisoner had a .45-caliber pistol that he wrestled from one of the military police. I ran from my office to the brig and there found a standoff. The prisoner was pointing the .45-caliber pistol at the guards and the guards were pointing their rifles at the escaped prisoner. Just like you see in the movies, I walked calmly toward the escapee saying "Just give me the gun. Everything will work out. It will be OK." But it did not work like the movies. As I neared the escapee, I held out my hand thinking he would hand me the gun. Instead, he jammed the gun into my stomach and pulled the trigger.

I was thrown forward as my father stepped on the brakes. His big hand was immediately upon me, keeping me from bumping my head on the dash board. My dad had been driving the old Plymouth station wagon slowly, but I had not been watching and several white-tailed deer had darted across the dirt road into our headlights. I realized my mouth was open, and now I was confused. My dad was here telling the story. Did he get shot? What happened?

Hearing this story was just part of a springtime ritual that I shared with my father. This ritual would begin at the morning breakfast table where he would say something like "I hear they are catching them up on the Tionesta. You have any plans after school?" And I would anticipate the trip all day long. My dad, who was self-employed, would leave work at 3:15 pm and we would travel to one of the local streams in search of trout. We would fish until dark, and on the way home he would tell me stories about his life. Most often, I would hear about his impoverished upbringing in a central Pennsylvania coal mining town and his life as a soldier in World War II and the Korean War. The stories always had a moral of some type. For example, a common theme drawn from his poverty-stricken youth was something like "That is why it is so important to get a good education. That is something they can never take away from you!" As a first-generation Italian who had been brought up during the Great Depression, my father knew something about having things taken from him. We would return home after dark and my mother would be ready to serve us the dinner that she had kept warm. My father and I would regale her with stories about the fishing trip to which she listened patiently.

It was through experiences like this that my parents shaped my thinking, my values, and my love for the outdoors. The origins of this book reside in the support I received from them and I dedicate it to them. There have never been two better parents or two better people.

Lt. Colonel Louis J. Manfredo received the Bronze Star (one of two) for the bravery he exhibited the day a prisoner escaped from the brig during the Korean War. The prisoner attempted to fire upon my father, but the gun jammed.

Preface

At the inception, the purpose of this book was straightforward. I hoped to provide an overview of the social psychological theories used in studying the human dimensions of wildlife management (HDW) and to suggest how the research guided by these theories can inform conservation practice. As I undertook the task, I assumed that I would address a "state of the science" in HDW. However, as I engaged in preparation for the book, I became increasingly aware that the recent advancements of the social sciences offer exciting new ways to explore human–wildlife relationships. As a result, the book is as much a reflection of my excitement for exploring new theoretical ideas as it is a report on the concepts of past HDW research. The reader will note four basic themes that frame this effort.

The first theme embraces the continued application of cognitive-based research in HDW. This represents the primary tradition of work in HDW and is addressed in the core topics of the book including attitudes, values, norms, and wildlife value orientations. The cognitive approach focuses primarily on what is learned and the deliberative thought by which people form evaluations of the world around them. It typically uses methods that elicit people's self-reports of their thoughts. Responses are analyzed and presented in a way that characterizes a population of interest or explores a hypothesized relationship. For a number of theoretical and practical reasons, this approach has played and will continue to play a central role in HDW. At the same time, it is important to recognize that we can broaden and deepen our understanding of human response to wildlife by looking beyond the cognitive aspects of human evaluation.

One direction for an expanded view is the growing attention given to the biological and evolutionary basis of thought. Trends of the mid-to-late twentieth century de-emphasized the role of genetic explanations of human behavior in part as a reaction to the support they gave repugnant notions of racial superiority. Yet there is now a resurgent interest in this topic. On one hand, findings suggest there may be human universals in response to wildlife. This might include, for example, the tendency to anthropomorphize, or assign human characteristics to wildlife. In addition, however, differences in human attitudes and behavioral patterns have a partial basis in inherited traits. For

example, there does, in fact, appear to be a genetic basis for a person's participation in recreational hunting (that combines with upbringing and environmental opportunity). But this is not a remnant of human aggressive tendencies;
it appears that there is also a genetic link to selection of all forms of occupation
and leisure.

Another theme that calls for attention in HDW is the realm of the feeling
states of individuals, i.e., human affect. The predominance of the cognitive and
rationalist tradition minimized the role of concepts like emotions and moods.
Good decision making, we have been taught, should eliminate emotions. Yet we
are now learning that emotion is a vital component of sound decision making.
Emotion is a critical concept for understanding our most fundamental, inherited responses to wildlife and for understanding the human experience that
results from wildlife encounters. I remain struck by Elster's (1999, p. 403) simple
but powerful quote "Emotions matter because if we did not have them nothing
else would matter."

Finally, there is growing interest in interdisciplinary approaches to developing theory. Prior reductionist approaches urged the social sciences to seek
simple cause–effect relationships, but increasingly this is regarded as an over-
simplistic representation of real-world problems. These are problems that
involve complex interactions among people and their environment. New
approaches try to model these complex phenomena and extend explanations
across multiple scales. I offer a multi-scale investigation of human response to
wildlife as a case study application in the last chapter of the book. To me, this
area offers an exciting frontier for the HDW scientist. I find it particularly
hopeful because it offers a venue for collaborative biological and social science
approaches in dealing with human–wildlife problems.

Within these four themes I see potential for advancements in our understanding of human–wildlife relationships. These are advancements that change
how we think about wildlife management problems, that can direct the strategies we undertake, and that can improve our conservation decisions. It is
certainly my wish that this book brings us closer to these advancements.

Fort Collins, USA   Michael J. Manfredo

Acknowledgments

The idea for this book hatched when I met with colleagues in a coffee house in Anchorage, Alaska. Several years later, a repeat of the Alaska experience provided an occasion for these colleagues to review the book's initial draft.

My first obligation is to thank these folks for taking their time and intellectual energy to help guide this book's development: Alan Bright, Peter Fix, David Fulton, Holly Stinchfield, Tara Teel, Jerry Vaske, and Harry Zinn. I suspect I would have never have made it to final drafts without reviews and input from Tara Teel, review by Peter Newman, and the varied contributions of Alia Dietsch. I know I would never have made it to final draft without Jamie Davis, a superb and tireless editor. Thanks also to the fine efforts of Ashley Dyer who edited earlier drafts of chapters in the book and helped in its initial organization.

On a personal note, thanks to my daughter, Anne, whose phone calls evoke an occasionally need to laugh. Thanks especially to my wife, Brenda, who endured seeing me in front of the computer at all hours of the day. Guess it is my turn to mow the lawn for the next 5 years.

Contents

Chapter 1
Who Cares About Wildlife?[1]

Contents

Introduction

A conversation I had aboard an airplane traveling from Houston to Denver inspired this book's title. After settling in my seat, I learned that the passenger beside me was an airline pilot. I asked him many questions during the flight, as air travel has always interested me. My newly made friend explained lift and drag, the workings of air traffic communications, and his preferred types of aircraft. I wanted to understand how he coped with the responsibility of so many lives. When he asked about my occupation – having just conversed about a profession recognized for its responsibility, social utility, and respect – I felt pressure to depict my work in a socially relevant context.

---

[1] Throughout this book I use the term "wildlife" in place of the phrase "fish and wildlife." It is for purposes of readability and not a deliberate exclusion of fishery issues. The book's topics are as applicable to fishery issues as to wildlife issues.

M.J. Manfredo, *Who Cares About Wildlife?*,
DOI: 10.1007/978-0-387-77040-6_1, © Springer Science+Business Media, LLC 2008

I described my current project. The project hoped to assess public attitudes toward a ballot initiative to ban wildlife trapping in Colorado. Wildlife management, as a profession, saw the impending ballot initiative as a threat. Managers felt that issues like wildlife trapping should be left to wildlife professionals and not voted on by the general public. Trapping, an important tool for Colorado's ranching community and an integral part of the pioneering spirit of rural Colorado, was also scrutinized by the public, a portion of whom were adamant because they viewed trapping as animal cruelty. The pilot's response surprised me: "Why do people get all upset about something like that? Who cares that much about wildlife anyway?" I considered that important question, and that question made me think of Jim, a wildlife manager who worked for the Colorado Division of Wildlife.

Jim devoted his life to his job, a job that demanded his attention at all hours. He enforced hunting regulations, educated the public about wildlife, and handled human–wildlife encounters that occurred in the urban fringe area. When I met Jim a mountain lion had just killed a teenage jogger, and Jim was dealing with a concerned public while also answering the questions of a grieving family.

Jim confessed he found the public unpredictable. Once when a mountain lion wandered into a residential area and settled onto a tree, the local TV station heard of the incident and arrived in time to film Jim and his co-workers shoot the lion with a tranquilizer. The lion tumbled roughly from the tree, and the whole scene was televised. That night, at home, Jim received anonymous phone calls; some of the callers threatened Jim's life due to his treatment of the lion.

Jim, the teenager's grieving family, the concerned public, and the anonymous callers all cared a great deal about wildlife.

People worldwide have different reasons for caring about wildlife: Wildlife are a source of attraction and fear, they have utilitarian value and symbolic meaning, they have religious or spiritual significance, and they are a barometer measuring people's concern for environmental sustainability. Four key areas of concern are:

- Their choice of recreation and tourism activities
- Their response to wildlife–human conflict
- Their interest in wildlife diseases
- Their concern for environmental sustainability

Wildlife-Associated Recreation and Tourism[2]

Leisure pursuits are increasingly important to people in post-industrial society, and they have a significant economic impact. The number of international

---

[2] Although some authors make distinction between tourism and recreation, I use the terms interchangeably here. The terms are used to denote purposive activity, typically including travel from home, for the purpose of enjoyment and rejuvenation.

tourists has doubled within the last decade (from the early 1990s to 2006). Globally, an estimated 842 million people were international tourists in 2006. The number of international tourists is expected to double again by 2020 (World Tourism Organization, 2006).

Tourism is the largest sector in the world economy generating $3.6 trillion in economic activity and 8% of jobs annually worldwide. Tourism is particularly important as an employer in poor regions of the world, and it is the primary export for 83% of developing countries (International Ecotourism Society, 2007).

Most forms of tourism that involve wildlife are classified as *nature-based tourism* or *ecotourism*. In the past 15 years, these forms of tourism have enjoyed significant growth relative to the rest of the tourism industry. In the 1990s ecotourism grew at a rate of 20–34% per year (Mastny, 2001), and in the twenty-first century, it continues to outpace the rest of the industry (International Ecotourism Society, 2007).

One way in which wildlife is important for tourism is as part of the entire package of an experience, for example, a side trip pursuit or a pleasant surprise while sightseeing. Analysis of North American ecotourism markets suggests that seeing wildlife is one of the top four setting attributes desired in a tourist experience (Wight, 1996). It would be impossible to gauge the full extent of the importance of wildlife in this support cast role for tourism.

A significant amount of tourism focuses on wildlife as the trip's primary purpose. Visitors to U.S. national parks rank viewing wildlife as a top reason for their attendance, and this would undoubtedly be true for visitors to a number of the world's protected areas, which now extend across 12% of the earth's surface (Brooks et al., 2004). Further evidence of wildlife as a driving force for tourism can be found in the 2006 U.S. Fish and Wildlife Service's (2007) *National Survey of Fishing, Hunting, and Wildlife-Associated Recreation*. This study showed that 87 million Americans participated in some form of wildlife-associated recreation, including hunting (12.5 million people), fishing (24.5 million people), wildlife viewing involving trips away from home (22.9 million people) or viewing in one's day-to-day residential life (77 million people). Other data hint at a widespread global interest in wildlife-associated recreation. For example, the 1996 *Survey on the Importance of Nature to Canadians* indicates that approximately 18% of the Canadian population, or 4.2 million people, participated in fishing, and 5% of the population, or 1.2 million people, participated in hunting (Federal-Provincial Task Force on the Importance of Nature to Canadians, 1999).

Participation trends show an interesting pattern. While hunting and fishing in the United States has declined, wildlife viewing has increased (Aiken, 1999; U.S. Fish and Wildlife Service, 2007). Simultaneously, many specialized forms of wildlife-associated recreation have grown considerably in the last two decades. Hoyt (2000) estimated that the number of whale-watching tourists worldwide increased from 4 million in 1991 to 9 million in 1998. Bucking the overall trend of hunting in the United States, participation in trophy hunting in

Africa has increased since the early 1990s, including a fourfold increase in Namibia and twofold increase in South Africa (Lindsey, Roulet, & Romanach, 2007). These few examples do not reveal the growth that is also occurring in the many specialty niche markets in wildlife-associated recreation including, for example, fly-fishing the flats in the tropics, bird viewing in Thailand or Costa Rica, viewing birds and reptiles on the Galapagos Islands, dolphin feeding in Australia, viewing monkeys at the temples in Singapore, and viewing butterflies in Mexico. The growth in these opportunities is related to strong consumer demand, the relatively low capital investment needed for ecotourism businesses, and the strong potential for local employment.

Wildlife-associated recreation generates a significant amount of economic activity. In 2006 U.S. hunters spent $23 billion, anglers spent $40 billion, and wildlife viewers spent $45 billion (U.S. Fish and Wildlife Service, 2007). Hoyt (2000) suggested that worldwide whale watching is a 1-billion-dollar industry operating in 492 communities and 87 countries and territories.

Locally the impact of wildlife-associated tourism can be quite significant. Aylward (2003), for example, reported that wildlife safari-centered nature tourism in the northeast Zululand part of the Kwazulu-Natal province South Africa accounts for 21% of gross geographic product and 30% of employment. Similarly, Navrud and Mungatana (1994) showed that the recreational value of wildlife viewing in Kenya was between $7.5 and $15 million. Orams (2000) reported that whale watching in the small South Pacific island community of Vava'u (population 16 thousand people) in Tonga yielded revenue in excess of $600,000 per year. Moreover, Andersson, Croné, Stage, and Stage (2005) examined gorilla tracking in Uganda and concluded tourist expenditures fall considerably short of willingness to pay. They suggested that the Bwinidi Impenetrable National Park could increase revenues sevenfold through additional fees.

Interest in wildlife is not restricted to those who take trips to the outdoors. The World Association of Zoos and Aquariums, with 1,200 organizations worldwide, estimates an attendance of 600 million people annually (World Association of Zoos and Aquariums, 2007). There has also been a growing market for TV and cinema that focuses on the natural habits of wildlife instead of prior programming that featured wildlife primarily as anthropomorphized characterizations of human life. *Animal Planet*, a TV channel launched in 1996 that televises features about animals (wild and domesticated), has experienced rapid growth. As of 2005 it reached 237 million subscribers in 160 countries who speak 24 languages (Broadband TV News, 2005). The interest and growth of this programming is global with the channel most recently expanding to Germany, Italy, and Vietnam (Discovery Communications Inc., 2004). Additional examples of the rapid growth are the rising numbers of subscribers: In Latin America the increase was by 24% between 1999 and 2000 to 9 million subscribers, and, likewise, in Asia the increase was by 205% to more than 24 million subscribers (BBC, 2001). Viewership of wildlife films is also expanding. *Winged Migration* (2003) and *March of the Penguins* (2005) grossed over $31 million and

$79 million worldwide respectively. These ticket sales make the films some of the most successful documentary films in history (Nash, 2005).

Growth of various forms of wildlife-associated tourism raises questions about the negative impacts that tourism and recreation may be having on wildlife populations. Recreation and tourism is in conflict with other deeply held public values such as concern for protection of wildlife and for environmental quality (addressed later in this section). An overview of the extent of these impacts is beyond the purpose of this chapter. See Knight and Gutzwiller (1995) for an introduction to the multiple ways that recreation can negatively impact wildlife species (e.g., direct effects through encounters, altering prey species, altering habitat, and habituation). Human–wildlife conflict is, however, an area of growing importance and I devote the next section to this topic.

Human-Wildlife Conflict

As I write this chapter, deer are in my backyard eating from my apple trees. In a few weeks, I will battle with the raccoons who always get to my corn just as it ripens. A woodpecker drums on the vent pipes of my house, telling everyone that the territory I regard as mine is also his. The morning paper reports that a jogger was bitten by a rattlesnake at a local reservoir. Ignoring the trailhead sign that warned of snakes, the jogger did not know the snake was dangerous and continued her run after being bitten. Fortunately, she encountered mountain bikers who immediately took her for help. Just last month an 11-year-old boy was killed by a bear while camping in Utah. The bear was believed to have become habituated to campers, finding campgrounds a good location to obtain easy food. For me, like many people in the world, undesirable wildlife encounters are part of daily life and their impacts can be significant.

Conover (2002) estimated that human–wildlife conflict causes $22.3 billion in losses per year in the United States alone. The largest expense ($8.3 billion) is incurred in urban areas and is due to mice, squirrels, raccoons, moles, pigeons, starlings, and skunks. Agricultural loss is estimated at $4.5 billion per year while annual loss due to deer–auto collisions is estimated at $1.6 billion.

Problems of human–wildlife conflict are not limited to the United States. Treves and Karanth (2004, p. 1492) noted this about human–carnivore conflicts:

This is a worldwide problem, exemplified by wolves (*Canis lupus*) and bears (*Ursus spp.*) that kill sheep in North America and Europe; Pumas (*Puma concolor*) and jaguars (*Panthera onca*) taking cattle in South America; numerous carnivore genera preying on cattle and goats in Africa; and tigers (*P. tigris*) and leopards (*P. pardus*) killing livestock in Asia.

The impact of this conflict is differentially distributed. Hill (2000) indicated that regionally aggregated impacts may not look significant, but for some individuals in high-conflict areas, impacts are devastating. Her research shows that some farmers in Uganda lost up to 60% of their crops to raiding by baboons.

In addition to economic and property loss, wildlife can threaten human safety. Conover (2002) claimed that attacks by alligators, cougars, bears, coyotes, bison, and moose have increased in the United States in recent decades. Such attacks are particularly problematic in rural areas of developing nations. Choudhury (2004) reported that in northeastern India, human–elephant conflict killed 1,500 people between 1980 and 2003. Retaliation and habitat loss have resulted in declining elephant populations, and in one area – Cachar, Assam – elephants have been extirpated. Rajpurohit and Krausman (2000) offered further evidence of the severity of impacts to humans; they reported that during a 6-year period (from 1989 to 1995) in south Bihar (India), elephants killed 242 people, sloth bears killed 50 people, and wolves killed 92 people.

Reasons explaining the increase in human–wildlife conflict vary. Expanding human settlement is believed to be the most critical reason: Driven by population pressures, economic growth, and the expanding global demand for natural resources, humans occupy more and more places. As this occurs, it destroys or fragments habitat, forcing humans and wildlife into confrontation. Because humans and wildlife share habitat, human–wildlife conflict is often coincidental; however, sometimes conflict occurs because humans want to be close to wildlife. For example, results show that 55.5 million Americans engage in wildlife feeding activities (U.S. Fish and Wildlife Service, 2007). Wildlife attracts tourists, and feeding wildlife has, in some cases, become a planned tourism activity (Orams, 2002). Feeding wildlife can lead to habituation, and animals may identify humans as a food source instead of a threat. Habituation is believed to be a factor in the death of a 9-year-old boy at an Australian World Heritage site; the boy was attacked by dingoes habituated through tourist feeding behavior (Thompson, Shirreffs, & McPhail, 2003). Beyond such direct attacks, human contact with wildlife can affect disease transmission. I review that concern in the next section.

Wildlife Disease

According to Wolfe, Dunavan, and Diamond (2007) the most important diseases of modern human populations have animal origins. These diseases emerged in the past 11,000 years following the rise of agriculture. Zoonotic (spread from animals to humans) diseases attract a great deal of attention and concern among researchers (Friend, 2006). Enserink (2000), citing findings of researchers from University of Edinburgh, claimed 1,709 pathogens plague humanity, half of which are zoonotic. Moreover, among the 156 pathogens considered to be emerging diseases, 73% are zoonotic.

Zoonotic diseases have had dramatic effect on the course of history. They have clearly impacted the ways of modern life (Wolfe et al., 2007). For example, human immunodeficiency virus (HIV) causes a disease that has greatly altered the lives of most of the world's population, infecting 40 million people worldwide and causing the death of about 3 million people in 2006 (Joint U.N. Programme on HIV/AIDS, 2007). HIV may have been transmitted to human populations from chimpanzees. Evidence of transmission is found in Gao et al. (1999) who identified the similarity between HIV and SIV, a virus common in chimps. Wolfe et al. (2004) provided evidence that transmission of SIV is common. Wolfe et al. (2004) tested the blood of more than 1,000 people in Central Africa who had regular contact with non-human primates. Approximately 1% of the people sampled had SIV.

Increasingly, research shows that the transmission of infectious agents between humans and primate species occurs in a variety of contexts affecting a wide range of people (Jones-Engel et al., 2006). Research by Jones-Engel et al. (2006) in Southeast Asia explored the possibility of transmission of several pathogens from macaques to humans. Macaques have a special status in Buddhist and Hindu cultures. They tend to congregate around temples where residents, workers, and tourists come in close contact with them. Through interaction with these macaques, people can be exposed through bites, scratches, or mucosal splashes. In these situations, there is a risk from transmission of pathogens such as herpes B, simian virus 40, simian foamy virus, and other simian retroviruses. The transmission of pathogens occurs both ways. Research shows macaques risk exposure to influenza and measles from humans (Jones-Engel, Engel, Schillaci, Babo, & Froehlich, 2001).

Many other zoonotic diseases, not borne from non-human primates, have also recently drawn attention. This includes the tick-borne Lyme disease, West Nile virus, SARS, and the possibility of avian flu. Chronic wasting disease (CWD) has not been transferred to humans, but this remains a possibility. CWD is a prion-based disease that kills deer and elk, and it has a significant effect on these animal populations in the United States. Recreational hunters in the United States harvest and consume thousands of these animals annually; this increases concern about the possibility of transmission of the disease. Should transmission occur, it would significantly affect human health. It would also negatively affect the economy of rural areas by discouraging tourism, and, due to lost license sales, decrease funding for state fish and wildlife agencies (Needham, Vaske, & Manfredo, 2004).

Conditions of globalization may influence the emergence of zoonotic diseases (Chomel, Belotto, & Meslin, 2007). Climate change accelerates mutation, while easy airline travel and the growth of urbanization facilitate disease transmission (Friend, 2006). Wolfe et al. (2007) proposed the need for an early warning detection system. They concluded that

> Most major human infectious disease have animal origins, and we continue to be bombarded by novel animal pathogens.. ... [M]onitoring should focus on people with high levels of exposure to wild animals, such as hunters, butchers of wild game, wildlife veterinarians, workers in the wildlife trade, and zoo workers. (p. 283)

While wildlife can have negative impacts on humans, such as through conflict and disease, it also symbolizes high environmental quality and life quality.
In the next section, I overview this area of concern.

The Condition of Wildlife and the Environment

In twentieth-century North America, people protected wildlife for utilitarian
reasons. At that time, game scarcity, extirpation, and extinction were part of a
growing number of concerns over the resiliency of wildlife populations. Dramatic forest fires followed extensive clear-cutting in the Midwest. Erosion
threatened farm productivity, and timber extraction outstripped regenerative
capabilities. Mining and industrial pollution diminished the ability of lakes and
rivers to support aquatic life (Frederick & Sedjo, 1991). With growing human
populations and technological advances, such as the repeating rifle, many
wildlife populations were rapidly depleted or became extinct (Harrington,
1991). The conservation leaders of that time, including Gifford Pinchot and
Aldo Leopold, were guided by a desire to ensure "sustained yield" of natural
materials. The emerging tradition of wildlife management during this time was
to convert hunting from exploitation to cropping (Peyton, 2000), and the results
were immensely successful. The North American wildlife management profession – including its effective regulation of harvest, restoration of game populations, philosophy of science-based management, politically powerful
stakeholder lobby, and sustained funding base (i.e., hunter and angler license
fees) – must be considered one of the most notable conservation success stories.

During the latter third of the twentieth century, utilitarian interests and the
single-species game management focus of the wildlife profession in North
America began giving way to broader concerns for ecological integrity, biodiversity preservation, and environmental quality. Legislation, such as the
Endangered Species Act of 1973, symbolized concern for losses, even among
species that had no utilitarian purpose. Studies of that time period show that a
significant number of Americans were opposed to recreational hunting (Kellert,
1978; Shaw, 1977), and early research on public values shows a diversity of
values toward wildlife, including a significant emphasis on non-utilitarian views
regarding the resource (Kellert, 1980). In this values transition, wildlife, once
viewed as villainous creatures, came to symbolize the lost ecological integrity
desired by the public. An illustration of this was revealed in a recent meta-
analysis of studies since the 1970s. The study showed that people in the United
States currently favor wolves and wolf reintroduction (Williams, Ericsson, &
Heberlein, 2002), whereas in the early twentieth century, the species was targeted for extirpation throughout the country.

In the last chapter of this book, Tara Teel and I provide evidence of a broad-
based North American shift from domination to mutualism in wildlife value

orientations. As a result of this shift, attitudes toward wildlife are more protection-oriented and concern has increased for the care of individual animals.

Shifting concern for wildlife is probably associated with the late twentieth-century growth in environmental concern. Dunlap (2002) tracked environmental attitudes in North America since the 1970s and results showed an overall trend of increasing concern. This is a worldwide trend, given the high level of pro-environmental attitudes found in both developed and developing nations (Dunlap, 1994; Inglehart, 1997). At the start of the twenty-first century, the growing awareness of global warming and its environmental consequences deepened environmental concern. Recent polling by ABC News/Washington Post/Stanford University (2007) showed a significant proportion of Americans agree that warming is occurring (84%), and 82% indicated this problem is *somewhat*, *very*, or *extremely* important to them personally.

The plight of wildlife due to an eroding environment is highly symbolic of the plight facing humanity. An excellent example is the declining habitat of polar bears, highlighted in Al Gore's movie *Inconvenient Truth*. During their migration to find a diminishing prey base, these bears have been caught on islands of drifting ice or left to swim incredible distances in search of refuge on stable ice. Due to global warming, which has led to extensive glacial melting in the polar regions, such refuge diminishes. The analogy is clear: Humans are on an "island" that is eroding, and, without major changes, they will be left with a fruitless struggle.

The growing concern is merited. As described in the Millennium Ecosystem Assessment (2005) during the past century, the extinction rate among wildlife species is 1,000 times greater than that indicated in fossil record. The rapid collapse of the North Atlantic cod fishery suggests the possibility of similarly sudden, cataclysmic change in the future. The rate of extinction is expected to accelerate: Predictions suggest that extinction will occur at a rate ten-times greater than that of the past century. Species loss is part of broad-based ecosystemic change in which factors such as population growth, economic growth, and the social and political factors that encourage growth are driving habitat change, climate change, spread of invasive species, over-exploitation of resources, and pollution (Millennium Ecosystem Assessment, 2005).

In the midst of all these changes, an important shift has occurred in the composition of the work force, overall philosophical orientation, and sphere of influence of the wildlife profession. The traditional institutions of the profession have been slow to respond to the changing nature of concerns for wildlife and the environment. State-level fish and wildlife agencies in the United States, in particular, continue to prioritize issues of recreational hunting and fishing. Newly emerging non-governmental organizations and new types of academic programs (e.g., Conservation Biology) champion the broader concerns for global biodiversity. The focus of this movement is on global issues, and its approach is ecosystemic and integrative across disciplines. Funding comes from grassroots memberships and private donations. As an illustration, from 1994 to 2004 operating revenues of World Wildlife Fund doubled (WWF, 2005). Half of the revenue in 2004 was from contributions ($66 million), with $29 million from WWF members alone. The Humane Society of the United States (HSUS) has experienced similar growth (Center for Consumer Freedom, 2005). In 1970 HSUS had an annual budget of about $500,000. By 1994, HSUS's annual revenue grew to $22 million, and by 2003 that number increased to $123 million.

In this new era of conservation, discourse about concern for environmental quality has taken an important turn. Once cast in opposition to economic growth and the forces of capitalism, environments are now discussed in the context of the services they provide to humans (Millennium Ecosystem Assessment, 2005). The development of markets for these services – as illustrated in the cases of carbon, wetlands, and water – brings the principles of capitalism to environmental protection (e.g., Hamilton, Bayon, Turner, & Higgins, 2007). Another important trend is that the concern for environmental quality is increasingly linked to discussions about the need to deal with worldwide poverty (Millennium Ecosystem Assessment, 2005). Long-term solutions to environmental degradation must consider the current imbalance of environmental benefits and costs for people. Without economic advancement in impoverished areas, unsustainable uses of wildlife and natural resources will occur as individuals favor their own personal well-being against broader environmental concerns.

Conclusion

In summary, these four areas – recreation and tourism, human-wildlife conflict, wildlife disease, and environmental sustainability – capture many of the bases of concern for, or caring about, wildlife. This list is not exhaustive and is confined primarily to non-domesticated, non-human animals. The benefits people receive from pet ownership would demand an even lengthier discussion (e.g., see McNicholas et al., 2005). However, it should be clear from the previous sections that concern for wildlife is prominent in contemporary society.

In the next section, I will address the emergence of *human dimensions of wildlife management* (HDW), a field of study that applies the social sciences to examine human–wildlife relationships, and, in doing so, provides information that contributes to effective wildlife conservation efforts. The field emerged in response to a need to deal with the multiple, and often-conflicting, public concerns over management and uses of wildlife.

The Human Dimensions of Wildlife

It should be clear from reading the previous section that wildlife management involves understanding and dealing with people. However, the study of wildlife management has been firmly rooted in the biological disciplines. Integration of the social sciences into wildlife management has occurred, but slowly. In this section I provide a brief overview of this trend and explore the emergence of a human-dimensions approach along two somewhat-independent lines. One line is associated with the growth of the North American tradition of wildlife management, and the second line I associate with anthropology, geography, and the growing cross-cultural, multidisciplinary interest in understanding human–wildlife relationships.

Human Dimensions as Part of the Wildlife Management Tradition of North America

Since the origin of the North American wildlife profession in the early twentieth century, the public has been a concern to wildlife managers (Witter & Jahn, 1998). In these early times, "citizens were portrayed as lacking knowledge to contribute to visionary conservation efforts" (Witter & Jahn, 1998, p. 202). As noted by King in 1948, it was "becoming increasingly apparent that the knowledge and cooperation of the public is of fundamental importance in carrying out a well-rounded conservation program" (p. 9). Given this view, the primary objective of interaction with the public was to provide education in order to ease the "management bottleneck" of the citizenry (Huboda, 1948). Not surprisingly, some of the early HDW efforts focused on public relations and led to an early text on that topic by Gilbert in 1971.

The latter half of the century was a period during which "people problems" began to attract the attention of researchers and managers alike. One of the earliest and most enduring HDW research efforts was the U.S. Fish and Wildlife Service's National Survey of Fishing, Hunting, and Wildlife-Associated Recreation. This survey was first conducted in 1955 (U.S. Bureau of Sport Fisheries and Wildlife, 1955) and has been conducted at regular intervals since then (U.S. Fish and Wildlife Service, 2007). The purpose of this survey has remained relatively constant over time: to track Americans' wildlife-associated recreation participation and economic expenditures. It was not until 10 years after that first survey that HDW investigations began to measure the attitudes and socio-demographic characteristics of various publics, focusing particularly on hunters and anglers.

The social sciences as they were used in wildlife management was recognized as a growing field of research in a review by Hendee and Potter in 1971. These researchers identified several topics of importance for future research, including hunter satisfaction, non-consumptive uses of wildlife, characteristics of the hunter population, access and hunting opportunities, economic impacts and values of wildlife, and political and legal issues in wildlife management. Two years later, Hendee and Schoenfeld (1973) introduced the term "human dimensions of wildlife" at a session of the North American Wildlife and Natural Resources Conference. While individual studies using social science techniques had been reported prior to that meeting, this was the first time an entire session was devoted to the topic (Witter & Jahn, 1998).

As the many management implications of HDW research were articulated, demand for HDW research, and the human dimensions of natural resources more generally, grew. Impetus for research came primarily from funding provided by federal, state, and local government agencies seeking assistance with the growing number of "people problems." The U.S. Forest Service Experiment Station was particularly important in fueling much of this early work. While the research initially focused on people's recreational uses of natural resources (including hunting and fishing), it became apparent that people's recreational interests were a subset of a broader array of natural resource-related topics involving the public (e.g., clear-cutting, water uses, and environmentalism). Researchers – primarily from rural sociology, agricultural economics, and recreation and parks disciplines – were leaders in providing momentum to this new area of human dimensions inquiry.

In the early 1980s, the Human Dimensions of Wildlife Study Group was formed and consisted of a small group of researchers who met on an informal basis to discuss their HDW research. From this beginning, a much larger international network of scientists with an interest in HDW began to emerge. This network expanded significantly in just the last decade of the twentieth century, during which time, outlets for scholarly work in HDW increased greatly.

The journal *Human Dimensions of Wildlife* was introduced in 1996, and this journal, along with traditional outlets for wildlife management-related research – including *Fisheries, Journal of North American Fisheries, Journal of Wildlife Management,* and the *Wildlife Society Bulletin* – became the stage for advancing HDW research. In addition, a variety of publications emphasizing broader natural resource and conservation-related topics – such as *Human Ecology Review, Conservation Biology, Environment and Behavior,* and *Society and Natural Resources* – have increasingly provided outlets for HDW work.

Since its introduction, HDW research has been primarily descriptive and applied. Its main focus is to provide information about public values that managers can consider while making wildlife decisions. Decker, Brown, and Siemer (2001, p. 3–4) emphasized the importance of this information:

> Wildlife management is based on human values. It exists because wildlife are viewed as a resource for people. When landowners practice management on their own lands, it reflects their personal values. When a state agency undertakes management on behalf of its citizens, it reflects community or social values in that state. North Americans' view of wildlife – our belief in their value for us – motivates wildlife management at all levels.

There is a *values* and *valuing* component in virtually all areas of wildlife management. For example, this valuing component is integral to managers when they: set regulations to control human use and taking of wildlife, enforce wildlife laws, educate the public about wildlife, ensure people's safety in relation to wildlife, control human activity in order to protect or enhance wildlife populations, balance human and economic well being with the health of wildlife populations, engage various publics in decisions about wildlife, manage agency staff, and work with legislators and other politicians.

Relative to the commitment to using biology in making management decisions, the growth of a HDW emphasis within wildlife agencies has been slow; however, interest continues to expand. As I write this in 2007, virtually all U.S. state fish and wildlife agencies sponsor HDW research on a regular or semi-regular basis and many agencies have HDW staff positions. To facilitate growth in this area, many universities have hired faculty with human-dimensions expertise who address issues in wildlife and natural resources more broadly. This has created new coursework and degree programs. Robertson and Butler (2001) identified 25 academic programs that offered human dimensions of natural resources coursework, most of which have been introduced in the last 15 years. Training opportunities are also expanding for existing professionals. The Western Association of Fish and Wildlife Agencies, for example, recently developed an HDW accreditation program for current employees within their agencies. It seems likely that this trend toward growth in training opportunities, expertise, and application of human dimensions in wildlife management will continue.

Broadening the Interest in Human Dimensions of Wildlife

The second area contributing to the emergence of a HDW tradition comes from sectors of anthropology and cultural geography. Because thoughts about wildlife are so prominent in pre-industrial societies, it would be logical that the topic of human–wildlife relationships has long been a concern of researchers in these fields. Exploration of such relationships early on was intended to help attain a theoretical understanding of basic topics like human social organization and evolution of human cognitive abilities. An explanation for the universal appearance of totemism in hunter-gatherer societies (see Chapter 7), for example, has been an enduring subject of debate among anthropologists (Willis, 1990). Understanding human–wildlife relationships, as a theme deserving its own area of study, has been a more recent focus. Shanklin (1985) concluded her review of human–animal relationships in *Annual Review of Anthropology* by stating "the investigation of human–animal relationships may well be one of the most fruitful endeavors in anthropology" (p. 380). The journal *Society and Animals,* introduced in the late 1990s, is providing an outlet for these recent endeavors.

While earlier studies in anthropology had a theoretical focus, more recent studies have taken on a highly applied emphasis (e.g., Knight, 2004) that is consistent with a call for anthropology to become more applied and involved in environmental issues (Milton, 1996). With a tradition of field work in pre-industrial or developing societies and the co-location of these societies in critical areas of high biodiversity value, anthropology and geography are well suited to engage in analyses of social conditions that can inform conservation action. Little (1999) declared the applied involvement of anthropology in environmental issues its own field of study. This field has explored the rise of environmentalism as a social movement, issues of indigenous rights, and the impacts on parks and protected areas with regard to such topics as poverty, development, social structure, and location of power and self-governance. The field also examines ecotourism. Ecotourism is a form of economic growth that is consistent with conservation goals, yet it often creates undesirable impacts on indigenous peoples (Little, 1999). Research on human–wildlife relationships in this applied area of anthropology includes topics related to the nature of human–wildlife conflict, wildlife damage compensation schemes, illegal trade of wildlife, co-management, subsistence issues, and the influence of global markets and policies on resource harvest. Another important trend from the anthropological line of research has been cross-disciplinary, integrative investigations (e.g., Galvin, Thorton, Roque de Pinho, Sunderland, & Boone, 2006). These studies model the complex interaction of social and biological forces to predict future conditions. These studies emerged from the cultural ecology and ecosystems traditions within anthropology and geography.

I will now contrast the anthropological tradition of exploring human–wildlife relationships (AT) with the HDW tradition associated with the wildlife profession in North America (NAT). AT has an international, cross-cultural emphasis, versus the tighter geographic focus of NAT. AT conducts work in association with NGOs, foundations, and development banks, while NAT's work is funded by governmental agencies charged with a more narrowly defined wildlife management mission. AT is more likely to employ qualitative research methods, whereas NAT is more likely to emphasize quantitative techniques. AT is more likely to examine issues pertaining to both domesticated and wild animal populations, while NAT focuses primarily on wild animal populations. Excluding the field of economics, AT's research is rooted in theory from anthropology and geography, whereas NAT's is based largely on theory from social psychology and sociology. AT is more likely to be engaged in cross-disciplinary research than NAT. AT often articulates an advocacy mission (conservation or animal rights) that is less apparent in NAT. Finally, aside from specialty conferences, the natural resources-oriented professional meetings for AT include those of societies like Society for Conservation Biology, The Society for Human Ecology, and the Ecological Society of America, while the corresponding organizations for NAT include The Wildlife Society, the American Fisheries Society, and the International Association for Society and Natural Resources, which is linked to the International Symposium for Society and Resource Management. I realize that the broad categorizations listed above have exclusions (e.g., economics, political science, and conservation psychology) and inevitably have individual exceptions. Regardless of their differences, these two traditions are contributing to an important understanding of human–wildlife relationships and are also helping practitioners deal with day-to-day wildlife conservation and management issues.

Why a Human Dimensions Approach to Wildlife Conservation and Management?

Deserved or not, the social sciences often have to justify their role in conservation and in science more generally. For example, in 2006, Senator Kay Bailey Hutchinson, Chair of the committee that provides oversight for the National Science Foundation (NSF), raised serious concerns about NSF funds being used for social science research (Mervis, 2006). Despite rebuttal (Ojima et al., 2006) and continuation of NSF funding for social science research, Hutchinson's concern merits response. Below, I offer a few of the primary reasons for including human dimensions considerations in wildlife management.

A Professional Imperative

Those who get involved in the wildlife profession typically do so because they have a deep passion for protection of natural resources. From that passion is born dedication, commitment, and a sense of ownership and priority in natural resource decisions. A fusion of knowledge about the resource with people own values can lead them to believe they can identify the "right" decision. When people believe they have the correct answer, they see their job as a need to convince others of the belief. This is a tricky business because, while others may benefit from knowing more about the resource, they may not share the manager's values and might differ on what are acceptable impacts or desired futures.

It is important to remember whose resource is at stake in natural resource decisions. In many cases, wildlife managers deal with governmentally owned resources that are managed by government agencies. The North American model of natural resource management, to which many of these agencies subscribe, was established on the foundation of the public trust doctrine. This doctrine, which can be traced to Roman law, proposes resources common to humans – including air; running water; the sea; and, in North America, wildlife – should be held in trust for all people of the state. While the legality of public ownership of wildlife resources is debated (Bean & Rowland, 1997), state wildlife agencies are heavily influenced by the public trust philosophy (Prukop & Regan, 2005; Western Association of Fish and Wildlife Agencies, 2006). The doctrine is also often discussed in a more global context in relation to the management of natural resources (Sand, 2004).

Under the trust doctrine, it is a central responsibility of the natural resource professional to serve the public interest, and that involves determining the public's values. The social sciences can help managers determine the public's values and can assist managers with functions associated with their trustee role, such as educating, representing, facilitating direct involvement of, and leading the public in arriving at decisions.

A Moral Imperative

Decisions about natural resources can have profound effects on the well-being of human populations, and there is an implied moral obligation to consider these impacts. For example, protected areas serve as places of refuge for large populations of wildlife that do not recognize the socio-political boundaries of parks. When the cultivated areas next to these parks offer an irresistible source of food, animals may partake and leave human families and communities in a greatly impoverished condition. While there may be no legal obligation in these situations, there is a moral obligation to make the impacts to these families part of our wildlife management plans.

We Can Learn About Our Constituents

The area of study where I have spent much of my professional career deals with understanding stakeholder attitudes and values toward natural resources. Across the many studies that I have conducted, the managers I work with are frequently surprised to learn about the views of the stakeholders they serve. For example, I conducted a study for the Colorado Division of Wildlife (CDOW) in the early 1990s to examine public values toward wildlife (Bright, Manfredo, & Fulton, 2000). Findings had implications for CDOW's focus on regulating hunting and fishing in the state. Sales of hunting and fishing licenses are particularly important to CDOW because they provide a substantial portion of the annual operating budget of the agency. CDOW managers were surprised to learn from our study that about three out of ten citizens in Colorado did not support recreational hunting. How, you might ask, could these wildlife managers be unaware that so many people felt this way? In part, it is because our impression of others is formed largely through our interactions with those we frequently come in contact with. Managers are more likely to come in contact with the most vocal stakeholders or those who are directly affected by managers decisions, for example hunters and anglers. Past studies document this tendency and show that managers are often poor judges of the opinions of the general publics the managers serve (e.g., Gigliotti & Harmoning, 2004). This fact underscores the importance of actively engaging in representative assessment and inclusion of publics in natural resource decisions.

We Can Learn from Our Constituents

Most wildlife and natural resource policy processes, both governmental and non-governmental, can be viewed as a variant of the comprehensive-rational model of decision making. At the most generic level, this approach proposes a process in which goals are set, alternative means of reaching them are evaluated, and an alternative is selected based on an analysis of consequences. The preferred alternative is then implemented, monitored, and evaluated. While there are many positive aspects of this model, one of its criticisms is its emphasis upon extensive scientific information provided through studies or by the manager who is trained in science. This heavy emphasis on data generation can be seen in early proposals on wildlife planning by pioneers of wildlife management such as King (1938). The fallacy of this approach has been the impossibility of obtaining complete information about the ramifications of a management decision. It also sets up unrealistic expectations that the manager could be the authoritative source of such extensive knowledge or expertise. As noted in Bailey, Elder, and McKinney's edited volume on wildlife conservation (1974, p. 577), "College training of wildlife biologists often emphasizes theory and the rational-comprehensive method of decision making. Yet practicing wildlife managers rely heavily on local experience providing empirical knowledge of habitats and populations." Increasingly, part of that local knowledge comes directly from stakeholders.

Recognizing the constraints of traditional decision models, there is increasing acceptance of the notion, particularly at the cross-cultural level, that we can learn a great deal from local and indigenous knowledge systems for purposes of wildlife and natural resources management.

There are a growing number of examples of this approach, including:

- Relying on the knowledge of local Belizeans to identify locations of spawning aggregations of fish (Drew, 2005).
- Learning from the practices of the Kayapo, who preserve corridors of mature forest between plots as a kind of biological reserve (Posey, 1988).
- Drawing upon the beliefs of rural residents of the western United States, who, in the 1980s, were approached by the Bureau of Land Management to identify areas of that meet criteria for wilderness classification.

Increasingly we recognize the importance of local knowledge in medicine, agriculture, wildlife management, and in addressing environmental issues. This adds an important task to natural resource management: developing approaches by which managers can learn about key types of information from local populations.

The Benefits of Investing in Social Capital

Researcher Ron Inglehart (1997) proposed that the latter half of the twentieth century witnessed a rising distrust of government as many developed countries progressed through phases of modernization and a rise in economic well-being. Governments forged to meet materialist needs (for example, those needs focused on physical and economic security) were not well suited to meet the growing diversity of interests of a post-modern society. The distrust of government is readily observable in natural resource management in the United States. It is marked by an increase in legal challenges to the actions of natural resource agencies and an explosion of citizen action groups and non-governmental organizations formed to affect natural resource decisions.

Most natural resource managers today realize they must deal with this lingering atmosphere of distrust. The notion of social capital accounts for the importance of trust in effective conservation action.

According to Cohen and Prusak (2001, p. 4):

> Social capital consists of the stock of active connections among people: the trust, mutual understanding, and shared values and behaviors that bind the members of human networks and communities and make cooperative action possible.

Pretty (2003) suggested that when social capital is high in formalized groups, people have the confidence to invest in collective activities, knowing that others will also invest. He reported on the growth of such collective efforts in dealing with issues associated with the management of watersheds, forests, irrigation, pest species, wildlife, and fisheries. These efforts, he suggested, offer a promising way to achieve sustainable management and governance of common resources.

Concepts of collaborative decision making, co-management, and participatory government are all directed toward improvement of social capital while achieving conservation goals (Child, 1996; Wondolleck & Yaffee, 2000). The efforts we take toward improving our understanding of the social dimension of wildlife management provide steps toward improving social capital; improvements in social capital, in turn, will improve our ability to achieve conservation goals.

Policy and Managerial Decisions Can Be Improved

Perhaps the most obvious reason for including a human dimensions perspective is that it can improve wildlife decisions. Better decisions are those more likely to reach their objectives, to endure over time, and to create the benefits we desire. Note that it is not guaranteed that social science information will make decisions easier. In many cases, information about the diversity of public interests will only clarify the potential impacts of decisions and differences among stakeholder groups with respect to their preferences and tolerances. It will not ensure consensus or eliminate controversy; it will, however, help to anticipate and define the nature of problems and guide action in dealing with conflicts.

The way in which the social sciences can inform decisions is multifaceted. For example, educational efforts will be more effective when they target what people already believe; tourism development will be more effective when it considers consumer demand; policy acceptance will be more likely when people feel they have been heard and can influence the decision; species conservation actions will be more enduring when they consider the impacts to local communities; resolving conflict among opposing stakeholder groups can benefit from an understanding of the basis for the conflict; effective representation of multiple values can occur through the use of collaborative models of decision making; and the provision of environmental services can be facilitated by an understanding of economic value and the types of socially acceptable techniques for securing that value. These and many, many more situations in wildlife and natural resources management will benefit from the application of social science techniques.

The Social Sciences Will Be Key to Creating Long-Term Conservation Solutions

The future of wildlife and areas of conservation must eventually be considered in the context of a broader array of contemporary global challenges. Issues such as poverty, disease, population growth, economic growth and disparity, and global warming will profoundly influence our ability to achieve conservation goals. These are largely social phenomena, and our ability to manage them depends in part on our ability to understand their social effects.

The world's population is now 6.5 billion people and the United Nations expects it to reach 9 billion people in just 35 years (Population Division of the Department of Economic and Social Affairs of the United Nations Secretariat, 2007). Rapidly modernizing countries, such as China and India, and expanding global economies are placing incredible pressure on the world's natural resources (Harris, 2003). Global climate change looms over all plans for the future. Despite an increasing awareness of consequences, the trajectory of climate change seems to be intensifying. China is expected to surpass the United States in carbon emissions by 2009 (Energy Information Administration, 2006), and these emissions will continue to increase. Between 2000 and 2020, the Chinese government expects a quadrupling of the country's GDP, and it has already been exceeding expectations toward that goal.

As our atmosphere warms, change will be extensive. The Intergovernmental Panel on Climate Change (Parry, Canziani, Palutikof, van der Linden, & Hanson, 2007) recently estimated that if temperatures rise by just 2–4°F, one-third of the world's species will be lost from their current range; they will either migrate elsewhere to escape rising temperatures or simply vanish. What demographic shift might accompany the shift in habitable lands and lands available for agriculture? What will be the status of protected areas, and how effective will we be in preserving biodiversity? At what point will changing public needs make politicians look to protected areas to solve issues of scarcity of agricultural land, habitable land, and essential commodities? These are, indeed, difficult challenges for which, I believe, we desperately need the involvement of the social sciences. To illustrate, what we learn through the social sciences can help us: develop effective ways of communicating with and affecting the behavior of publics; determine ways to engage in effective action for the common good; develop political structures that can react more effectively to pressing environmental threats; understand the various types of services provided by biodiversity and healthy, functioning ecosystems, and help us establish mechanisms for sustaining these services; understand and avoid negative social impacts from resource uses and resource policies across multiple geographic scales; and, with the help of ecological and biophysical knowledge, predict future occurrences under different scenarios.

Why This Book?

The category *social sciences* embraces many disciplines and applied areas of study. In this book, I focus on just a small portion of these disciplinary perspectives, including a sample of theories from social psychology. As societal concern for the environment has expanded, there have been recent calls for psychology to become more involved in natural resource issues (Clayton & Brook, 2005; Saunders, 2003). Social psychology offers the promise of understanding, predicting, and affecting human thought and behavior toward wildlife in ways that can improve our ability to achieve conservation goals.

The overview of concepts provided in this book is not exhaustive; it builds upon concepts that have, for four decades, guided research in exploring the human dimensions of wildlife and natural resources. In an effort to expand the context and utility of these cognitive concepts, topics of emotions, heritability, and culture are also provided. The purpose of the book is (a) to provide an overview of the conceptual approaches currently used in studying human–wildlife relationships, (b) to stimulate the introduction of new approaches for examining such relationships, and (c) to suggest needed trends for future research in this area. The book is targeted at students of human–wildlife relationships and of the role of the social sciences in natural resources more generally. It is not limited to formal students; rather, it includes the many academics, practitioners, and future professionals who are attracted to this area of study.

Concepts for Examining Human–Wildlife Relationships

The structure of this book builds toward a multilevel model of human response to wildlife. At the *individual level* (micro level), human response to wildlife is seen as a learned response drawing upon a foundation of inherited tendencies such as anthropomorphizing. Human response to wildlife is based on a complex mix of emotions and cognitions. In the cognitive domain, human thought is viewed structurally as building from the basic (values) to specific (attitudes) with a strong influence from social group involvement (norms). At a *societal/cultural level,* the behavior and cognitive makeup of individuals has material and symbolic associations. The cognitive structure of people within a society is in an adaptive relationship with material factors such as economy, demography, and the environment.

Chapter 2 begins with an overview of genetically based tendencies that shape human response to wildlife. For the vast majority of human history, wildlife have been linked in some way to our survival needs including safety, security, shelter, and sustenance. It is highly likely that the nature of evolved human characteristics was shaped by the threats and opportunities provided by wildlife. Today, situations eliciting human responses of surprise and fear from wildlife encounters might be the most obvious examples of inherited responses to wildlife. However, it is quite likely that *most* human responses toward wildlife build upon a foundation of inherited tendencies. Co-author David Fulton and I explore research that suggests there are human universals that shape response to wildlife as well as the notion that heritability explains individual differences in human response to wildlife.

Chapter 3 introduces the concept of emotion, which is perhaps the most obvious bridge between genetic and learned explanations of human–wildlife relationships. While some theorists emphasize and define emotion by physiological reactions and responses, others emphasize the role of cognition and comprehension, or the role of culture in fashioning emotion. Emotion and related topics, such as mood and affect, have been overlooked as a topic of study during much of the twentieth century because they were seen as a deterrent to rational thought. More recently, research shows that emotion is an essential part of sound decision making. It also affects group processes and conflict resolution, the storage and recall of memories, and persuasion and attitude change. While emotion theorists often use wildlife–human interaction situations to illustrate emotional response (for example, a bear chasing a human), research examining human emotional response to wildlife is sparse. The experience of encounters with wildlife, people's attractions and interest in wildlife, and the intensity of conflict over wildlife are imbued with varying types and degrees of emotion. If we are to understand the individual meanings of wildlife to people, it is essential to attain a better understanding of emotion in human–wildlife relationships.

Chapter 4 (Attitudes), which is co-authored with Alan Bright, Chapter 5 (Norms), and Chapter 6 (Values) present the core concepts of the cognitive hierarchy that represents the learned component of human response to wildlife. Social scientists use the term *cognitions* to refer to thoughts based in learning and socialization. The cognitive realm is represented as a hierarchy of interconnected concepts including values, value orientations, norms, and attitudes. Most of the research that has been conducted in HDW draws, either directly or indirectly, from cognitive concepts.

Chapter 7 explores questions about human–wildlife relationships that emerge in a cross-cultural perspective. Are there predictable ways that human––wildlife relationships are structured by stages of cultural development, cultural organization, or religious orientation? Why is the custom of totemism so critical in understanding human social organization and cognitive tendencies? In addition, this chapter explores concepts that model cultural shift, which helps us examine a trend of shifting wildlife value orientations in post-industrial society. The latter topic is the concern of the last chapter, co-authored with Tara Teel (Chapter 8).

Chapter 8 provides a demonstration of multilevel research for exploring the effect of modernization on value shift. It introduces a conceptual approach for integrating value orientations into the VAB hierarchy and defining value orientations as a reflection of cultural ideology. Chapter 8 also offers insight into important future directions for research on human–wildlife relationships.

Conclusion

In the mid-1990s, I was involved in a collaborative human dimensions partnership with a state-level agency responsible for managing fish and wildlife. While setting up the partnership, we held many planning meetings at which we discussed goals, purpose, and procedure. The agency and the university approached the partnership from different perspectives, which reflects the unfortunate trend of separation between universities and practitioners in the natural resources fields.

The agency, for example, was interested in straightforward information that could be attained quickly and could help the state's commissioners make decisions. The university partners were interested in student education, theory building, research opportunities, publication prospects, etc. While differences were apparent, we found an important commonality that served us well as we encountered the challenges of our work relationship. We agreed to strive toward making better decisions about wildlife. I hope that this book leads to better research, advances theory, and assists in educating of professionals, and that this ultimately translates into better decisions about wildlife.

Summary

- Humans care a great deal about relationships with wildlife. Wildlife are the focus of a significant amount of tourism and recreation that generates considerable economic impact. Human–wildlife conflict is also a concern. It threatens human safety and creates considerable economic costs. Wildlife are a source of many human diseases. Globalization and climate change may increase the threat from emerging zoonotic diseases. There is widespread concern for the health of wildlife populations. The health of these populations is part of a broad concern for environmental quality and sustainability.
- The field of study known as human dimensions of wildlife has emerged to provide information about stakeholder concerns and values about wildlife. This field of study has developed along two lines. The first was associated with the North American tradition of wildlife management, the second through long-standing interest of anthropology and geography in human–wildlife relationships. These two traditions differ in terms of methods used, geographic focus, level of advocacy, and disciplinary orientation.
- The human dimensions sciences can offer unique contributions to wildlife conservation. They can provide information that helps decision makers understand the interests of stakeholders and meet the public trust doctrine. They can direct attention to the social ramification of wildlife decisions and moral obligations related to negative impacts of policy decisions. They facilitate learning about stakeholders and repeatedly show we cannot trust our personal perceptions of public preference. They increase our awareness

that we can facilitate conservation goals through the inclusion of local indigenous knowledge. They show how engaging public builds social capital which facilitates effective decisions. Finally, use of the social sciences will lead to better (longer lasting, reach desired outcomes, less controversial) short- and long-term decisions.

- Among the many social science perspectives, this book focuses on a cognitive approach to examining human–wildlife relationships. The core of the cognitive approach – values, value orientations, attitudes, and norms – is extended by linking it to topics of heritability, emotions, and cultural-level forces of cognitive shift.
- The book proceeds on the assumption that improvements in application of social science concepts will lead to better social science information. That, in turn, will lead to better decisions about wildlife.

References

ABC News/Washington Post/Stanford University. (2007). *Concern soars about global arming as world's top environmental threat.* Press release found at http://www.eesi.org/briefings/ 2007/Energy%20&%20 Climate/5-4-07_Climate_polling/GW%202007%20ABC%20 News%20Release.pdf.

Aiken, R. (1999). 1980–1995 participation in fishing, hunting, and wildlife watching: National and regional demographic trends. Report 96-5. U. S. Fish and Wildlife Service. Division of Federal Aid, Washington, DC.

Andersson, P., Croné, S., Stage, J., & Stage, J. (2005). Potential monopoly rents from international wildlife tourism: An example from Uganda's gorilla tourism. *Eastern Africa Social Science Research Review, 21*(1), 1–18.

Aylward, B. (2003). The actual and potential contribution of nature tourism in Zululand: Considerations for development, equity and conservation, In B. Aylward, & E. Lutz (Eds.), *Nature tourism, conservation, and development in Kwazulu-Natal, South Africa.* Washington: World Bank.

Bailey, J. A., Elder, W., & McKinney, T. D. (Eds.). (1974). *Readings in wildlife conservation.* Washington: The Wildlife Society.

BBC. (2001). *BBC worldwide 2000/01.* Retrieved September 12, 2005, from http://www. bbcworldwide.com/review/channels2.html.

Bean, M. J., & Rowland, M. J. (1997). *The evolution of national wildlife law.* Westport, CT: Praeger Publishers.

Bright, A., Manfredo, M. J., & Fulton, D. (2000). Segmenting the public: An application of value orientations to wildlife planning in Colorado. *Wildlife Society Bulletin, 28*(1), 218–226.

Broadband TV News. (2005, April 15). *Animal Planet launches in Italy.* Retrieved September 12, 2005, from http://www.broadbandtvnews.com/archive_uk/.

Brooks, T. M., Bakarr, M. I., Boucher, T., DaFonseca, G. A., Hilton-Taylor, C., Hoekstra, J. M., et al. (2004). Coverage provided by the global protected-area system: Is it enough? *Bioscience, 54*(12), 1081–1091.

Center for Consumer Freedom. (2005). *Activist cash: Humane Society of the United States.* Retrieved September 12, 2005, from http://www.activistcash.com/ organization_overview. cfm/oid/136.

Child, B. (1996). The practice and principles of community-based wildlife management in Zimbabwe: the CAMPFIRE programme. *Biodiversity and Conservation, 5*(3), 369–398.

Chomel, B. B., Belotto, A., & Meslin, F. X. (2007). Wildlife, exotic pets, and emerging zoonoses. *Emerging Infectious Disease, 13*, 6–11.

Choudhury, A. (2004). Human-elephant conflicts in Northeast India. *Human Dimensions of Wildlife, 9*, 261–270.

Clayton, S., & Brooke, A. (2005). Can psychology help save the world? A model for conservation psychology. *Analysis of Social Issues and Public Policy, 5*(1), 87–102.

Cohen, D., & Prusak, L. (2001). In *good company: How social capital makes organizations work* (pp. 214 and xiii). Boston, Ma: Harvard Business School Press.

Conover, M. (2002). *Resolving human-wildlife conflicts: The science of wildlife damage management*. Washington: Lewis Publishers.

Decker, D. J., Brown, T. L. & Siemer, W. F. (2001). Evolution of people-wildlife relations. In D. J. Decker, T. L. Brown, & W. F. Siemer (Eds.), *Human Dimensions of Wildlife in North America* (pp. 3–22). Bethesda, MD: The Wildlife Society.

Discovery Communications Inc. (2004). *International networks: Discovery networks international*. Retrieved September 12, 2005, from http:// corporate.discovery.com/brands/ networks_abroad.html.

Drew, J. A. (2005). Use of traditional ecological knowledge in marine conservation. *Conservation Biology, 19* (4), 1286–1293.

Dunlap, R. E. (1994). International attitudes towards environment and development. In H. O. Bergesen, & G. Parmann (Eds.), *Green Globe Yearbook of International Co-operation in Environment and Development* (pp. 115–126). Oxford: Oxford University Press.

Dunlap, R. E. (2002, September/October). An enduring concern: Light stays green for environmental protection. *Policy Perspective, 13* (5), 10–14.

Energy Information Administration. (2006). *International energy outlook*. Office of Integrated Analysis and Forecasting, Department of Energy, Washington DC.

Enserink, M. (2000). Malaysian researchers trace Nipah virus outbreak to bats. *Science, 27*, 518–519.

Federal-Provincial Task Force on the Importance of Nature to Canadians. (1999). *The importance of nature to Canadians: Survey highlights*. Retrieved September 12, 2005, from http://www.ec.gc.ca/nature/highlite.html.

Frederick, K. D., & Sedjo, R. A. (Eds.). (1991). *America's renewable resources: Historical trends and current challenges*. Washington: Resources for the Future.

Friend, M. (2006). *Disease emergence and resurgence: The human-wildlife connection*. Reston, VA: U.S. Geological Survey, Circular 1285.

Galvin, K. A., Thorton, P. K., Roque de Pinho, J., Sunderland, J., & Boone, R. B. (2006). Integrated modeling and its potential for resolving conflicts between conservation and people in the rangelands of East Africa. *Human Ecology, 34*(2), 155–183.

Gao, F., Bailes, E., Robertson, D. L., Chen, Y., Rodenburg, C. M., Michael, F. S., et al. (1999). Origin of HIV-1 in the chimpanzee Pan troglodytes. *Nature, 397*, 436–444.

Gigliotti, L., & Harmoning, A. (2004). Findings abstract. *Human Dimensions of Wildlife, 9* (1), 79–81.

Gilbert, D. L. (1971). *Natural resources and public relations*. Bethesda, MD: The Wildlife Society.

Hamilton, K., Bayon, R., Turner, G., & Higgins, D. (2007). *State of the voluntary carbon markets 2007: picking up steam*. Washington, DC: The Katoomba Group's Ecosystem Marketplace.

Harrington, W. (1991). Wildlife: Severe decline and partial recovery. In K. D. Fredrick, & R. A. Sedjo (Eds.), *America's Renewable Resources: Historical trends and current challenges* (pp. 205–248). Washington: Resources for the Future.

Harris, P. G. (Ed.). (2003). *Global warming and East Asia: The domestic and international politics of climate change*. London: Routledge.

Hendee, J. C., & Potter, D. R. (1971). Human behavior and wildlife management: Needed research. *Transactions of the North American Wildlife and Natural Resources Conference, 36*, 383–396.

Hendee, J. C., & Schoenfeld, C. (1973). Human dimensions in wildlife programs. *Transactions of the North American Wildlife and Natural Resources Conference, 38*, 182.

Hill, C. M. (2000). A conflict of interest between people and baboons: Crop raiding in Uganda. *International Journal of Primatology, 21* (2), 299–315.

Hoyt, E. (2000). Whale-watching 2000: *Worldwide tourism numbers, expenditures, and expanding socioeconomic benefits.* Crowborough: International Fund for Animal Welfare.

Huboda, M. (1948). An uninformed public: The management bottleneck. *Transactions of the North American Wildlife and Natural Resources Conference, 13*, 141–142.

International Ecotourism Society. (2007). *Ecotourism Fact Sheet.* www.ecotourism.org.

Inglehart, R. (1997). *Modernization and postmodernization.* Princeton, NJ: Princeton University Press.

Joint U. N. Programme on HIV/AIDS. (2007). *Global Summary of the AIDS epidemic, December 2006.* http://data.unaids.org/pub/EpiReport/2006/02-Global_Summary_2006_EpiUpdate_eng.pdf.

Jones-Engel, L., Engel, G. A., Heidrich, J., Chalise, M., Poudel, N., Viscidi, R., et al. (2006). Temple monkeys and health implications of commensalism, Kathmandu, Nepal. *Emerging Infectious Diseases, 12*, 900–906.

Jones-Engel, L., Engel, G. A., Schillaci, M. A., Babo, R., & Froehlich, J. (2001). Detection of antibodies to selected human pathogens among wild and pet macaques (*Macaca tonkeana*) in Sulawesi, Indonesia. *American Journal of Primatology, 54*(3), 171–178.

Kellert, S. R. (1978). Attitudes and characteristics of hunters and anti-hunters. *Trans. North American Wildlife and Natural Resources Conference, 43*, 412–423.

Kellert, S. R. (1980). American attitudes toward and knowledge of animals: An update. *International Journal for the Study on Animal Problems, 1*(2), 87–112.

King, F. H. (1948). The management of man. *Wisconsin Conservation Bulletin, 13*(9), 9–11.

King, R. T. (1938). The wildlife management plan. Excerpt from The essentials of wildlife range. *Journal of Forestry, 36*(5), 457–464. Reprinted in J. A. Bailey, W. Elder, & T. D. McKinney (Eds.), (1974). *Readings in wildlife conservation, (*pp. 573–575). Washington: The Wildlife Society.

Knight, J. (Ed.). (2004). *Wildlife in Asia.* Routledge: Curzon, London.

Knight, R. L., & Gutzwiller, K. J. (Eds.). (1995). *Wildlife and recreationists.* Washington: Island Press.

Lindsey, P. A., Roulet, P. A., & Romanach, S. S. (2007). Economic and conservation significance of the trophy hunting industry in sub-Saharan Africa. *Biological Conservation, 134*, 455–469.

Little, P. E. (1999). Environments and environmentalism in anthropological research: facing a new millennium. *Annual Review of Anthropology, 28*, 253–284.

Mastny, L. (2001). Treading lightly: New paths for international tourism, *Worldwatch Paper, 159*, Washington, DC: Worldwatch Institute.

McNicholas, J., Gilbey, A., Rennie A., Ahmedzai, S., Dono, J., & Ormerod, E. (2005). Pet ownership and human health: a brief review of evidence and issues. *BMJ, 331*, 1252–1254.

Mervis, J. (2006). Senate panel chair asks why NSF funds social sciences. *Science, 312*(829), 1470.

Millennium Ecosystem Assessment. (2005). *Ecosystems and human well-being: Synthesis.* Washington, DC: Island Press.

Milton, K. (1996). *Environmentalism and cultural theory.* London: Routledge.

Nash, B. (2005). *The numbers: Box office history for documentary movies.* Retrieved September 14, 2005, from http://www.the-numbers.com.

Navrud, S., & E. Mungatana. (1994). Environmental valuation in developing countries: The recreational value of wildlife viewing. *Ecological Economics, 11*, 135–151.

Needham, M. D., Vaske, J. J., & Manfredo, M. J. (2004). Hunters' behavior and acceptance of management actions related to chronic wasting disease in eight states. *Human Dimensions of Wildlife, 9*, 211–231.

Ojima, D. S., Wall, D. H., Moore, J., Galvin, K., Hobbs, N. T., Hunt, W. H., et al. (2006). Don't sell social science short. *Science, 312,* 1470.

Orams, M. B. (2000). *The economic benefits of whale-watching in Vava'u, the Kingdom of Tonga.* New Zealand: Centre for Tourism Research, Massey University at Albany.

Orams, M. B. (2002). Feeding wildlife as a tourism attraction: a review of issues and impacts. *Tourism Management, 23*(3), 281–293.

Parry, M., Canziani, O., Palutikof, J., van der Linden, P., Hanson, C. (Eds.). (2007). Climate change 2007: Impacts, adaptation and vulnerability. Contribution of working group II to the *Fourth Assessment Report of the Intergovernmental Panel on Climate Change.* Cambridge: Cambridge University Press.

Peyton, R. B. (2000). Wildlife management: Cropping to manage or managing to crop? *Wildlife Society Bulletin, 28,* 774–779.

Population Division of the Department of Economic and Social Affairs of the United Nations Secretariat. (2007). *World Population Prospects: The 2006 Revision* and *World Urbanization Prospects: The 2005 Revision,* http://esa.un.org/unpp, Monday, October 01, 2007; 11:00:23 AM.

Posey, D. (1988). Kayapo'Indian natural-resource management. In J. S. Denslow, & C. Padoch (eds.), *People of the Tropical Rainforest.* Berkeley: University of California Press.

Pretty, J. (2003, December 12). Social capital and the collective management of resources. *Science, 302*(5652), 1912–1914.

Prukop, J., & Regan, R. J. (2005). The value of the North American model of wildlife conservation – an International Association of Fish and Wildlife Agencies position. *Wildlife Society Bulletin, 33*(1), 374–377.

Rajpurohit, K. S., & Krausman, P. R. (2000). Human-sloth-bear conflicts in Madhya Pradesh, India. *Wildlife Society Bulletin, 28*(2), 393–399.

Robertson, R., & Butler, M. J. (2001). Teaching human dimensions of fish and wildlife management in U.S. universities. *Human Dimensions of Wildlife, 6*(1), 67–76.

Sand, P. H. (2004). Sovereignty bounded: public trusteeship for common pool resources? *Global Environmental Politics, 4*(1), 47–71.

Saunders, C. (2003). The emerging field of conservation psychology. *Human Ecology Review, 10,* 137–153.

Shanklin, E. (1985). Sustenance and symbol: Anthropological studies of domesticated animals. *Annual Review of Anthropology, 14,* 375–403.

Shaw, W. (1977). A survey of hunting opponents. *Wildlife Society Bulletin, 5*(1), 19–24.

Thompson, J., Shirreffs, L., & McPhail, I. (2003). Dingoes on Fraser Island – Tourism dream or management nightmare? *Human Dimensions of Wildlife, 8*(1), 37–47.

Treves, A., & Karanth, U. (2004). Human-carnivore conflict and perspectives on carnivore management worldwide. *Conservation Biology, 17*(6), 1491–1499.

U.S. Bureau of Sport Fisheries and Wildlife. (1955). *National survey of fishing and hunting.* Washington, DC: U.S. Government Printing Office.

U.S. Fish and Wildlife Service. (2007). *2006 National survey of fishing, hunting, and wildlife-associated recreation: National overview.* Washington, DC: U.S. Fish and Wildlife Service.

Western Association of Fish and Wildlife Agencies. (2006). *Resolution: The public trust doctrine on fish and wildlife conservation.* Retrieved August, 2007 from http://montanatws.org/chapters/mt/pdfs/WAFWAResolution-PublicTrustDoctrine.pdf.

Williams, C. K., Ericsson, G., & Heberlein, T. A. (2002). A quantitative summary of attitudes toward wolves and their reintroduction (1972–2000). *Wildlife Society Bulletin, 30*(2), 575–584.

Willis, R. (1990). Introduction. In R. Willis (Ed.), *Signifying animals: human meaning in the natural world* (pp. 1–24). London: Routledge:.

Witter, D. J., & Jahn, L. R. (1998). Emergence of human dimensions in wildlife management. *Transactions of the 63rd North American and Natural Resources Conference, 63,* 200–214.

World Association of Zoos and Aquariums. (2007). Retrieved July 2007 from http://www.waza.org/network/index.php?main=zoos.

Wolfe, N. D., Dunavan, C. P., & Diamond, J. (2007). Origins of major human infectious diseases. *Nature, 447*, 279–283.

Wolfe, N. D., Switzer, W. M., Carr, J. K., Bhullar, V. B., Shanmugam, V., Tamoufe, H., et al. (2004). Naturally acquired simian retrovirus infections in Central African Hunters. *The Lancet, 363*, 932.

Wondolleck, S., & Yaffee, L. (2000). *Making collaboration work: Lessons from innovation in natural resource management*. Washington: Island Press.

World Tourism Organization. (2006). *Tourism highlights: 2006 edition*. Retrieved (date needed) from http://www.world-tourism.org/facts/menu.html.

World Wildlife Fund. (2005). *2004 annual report*. Retrieved September, 2005 from http://www.worldwildlife.org/ about/2004_report/financials.pdf.

Wight, P. (1996). North American ecotourism markets. *Journal of Travel Research, 35*(1), 3–10.

Chapter 2
The Biological Context of Wildlife Values: Are There Etchings on the Slate?[1]

Contents

> *A man may not care for golf and still be human, but the man who does not like to see, hunt, photograph, or otherwise outwit birds or animals is hardly normal. He is supercivilized, and I for one do not know how to deal with him. Babes do not tremble when they are shown a golf ball, but I should not like to own the boy whose hair does not lift his hat when he sees his first deer. We are dealing, therefore, with something that lies pretty deep.*
> Aldo Leopold, 1922

Introduction

In this chapter, we examine the genetic influences on human response to wildlife. Musings about the nature of humans may seem unlikely to provide direction to wildlife professionals in their day-to-day dealings with wildlife; however, our beliefs about human nature shape our most basic assumptions about the cause of human action. What part of our response to wildlife is dictated by what

---

[1] This chapter was co-authored with David C. Fulton.

M.J. Manfredo, *Who Cares About Wildlife?*,
DOI: 10.1007/978-0-387-77040-6_2, © Springer Science+Business Media, LLC 2008

we inherit and how we have evolved? Scientific thinking about the role of heritability has changed significantly over the past 25 years and has served to balance the mid-twentieth century view that most of our action and thought is shaped through cultural learning. For students of human–wildlife relationships, an examination of our biological foundations helps explore why our cognitive and emotional abilities have developed as they have. Clearly, wildlife are not merely incidental to this development. For example, authors such as Quammen (2003) and Hart and Sussman (2005) contend that the presence of predators influenced how humans developed both ecologically and psychologically. This chapter provides an introduction to the growing heritability literature as it relates to questions about human–wildlife relationships.

The chapter begins by reviewing perspectives on nature's role in shaping human response systems and is followed by a review of concepts regarding the biological basis of values toward wildlife.

A Short History of Biology and the Social Sciences

Most people who share a scientific perspective readily embrace the idea that humans are biological organisms that as a species have evolved over millennia and whose physical traits – such as sex, height, and eye color – are largely determined by genes that are biologically inherited. The degree to which complex human behaviors and psychological characteristics (e.g., personality, values, preferences, and culture) have biological bases, however, has been a contentious topic in the social sciences (McGee & Warms, 1996; Pinker, 2002). The most broadly held perspective in the social sciences is that complex behaviors and characteristics are learned through socialization and enculturation.

From this perspective, genetics and biological evolution play very little, if any, role in acquiring such learned characteristics. While we may have evolved as biological organisms, socialization and enculturation have replaced genetic factors as the dominant forces shaping the psychological and social characteristics of individuals. The often-used metaphor is that at birth humans are *tabla rasa* (a blank slate) on which the forces of socialization and enculturation act to largely determine who we are and what we think (Pinker, 2002). Thus, we are almost exclusively products of our social environments. This general view of social and cultural phenomena has been termed the Standard Social Science Model (Tooby & Cosmides, 1992).

While the fields of anthropology, psychology, and sociology were not much influenced by biological theory during the twentieth century, biological thinking and, particularly, Darwinian notions of evolution and adaptation were instrumental to the early formation of these fields (Turner et al., 1997). The early use of Darwinian notions of *natural selection* to support morally repugnant and erroneous ideas of Social Darwinism, racial superiority, and eugenics (e.g., Clossen, 1897), however, led to long-lasting suspicions of biological ideas in the social sciences. These suspicions intensified when specious biological arguments were used by Hitler and Nazi Germany to support racist policies and horrific crimes against humanity (Mazur & Robertson, 1972; Opler, 1945). Within psychology, the prevalence of Skinnerian behaviorism and its emphasis on the role of the external environment in shaping individual behaviors further discouraged the use of biological theories of evolution (Skinner, 1971; Turner et al., 1997).

While there are notable exceptions (e.g., ecological anthropology [Steward, Steward, & Murphy, 1977] and human ecology [Hawley, 1986]), biological theorizing played a minor role in the social sciences for most of the twentieth century. As noted by Mazur and Robertson (1972), the biological bases of human behavior had popular appeal in works such as Morris' (1967) *The Naked Ape* where it was used to explain the violent tendencies of humans (see also Ardrey, 1961; Ardrey, 1966; Tiger, 1969). Serious reconsideration of evolutionary thinking and biological ideas about humans, however, began with the work of Wilson (1975). In particular E. O. Wilson's extension of sociobiology to the evolution of human social behavior was a starting point for researches to re-examine the potential role of biological evolutionary theory in the social sciences (Dawkins, 1976). The advocates of sociobiology, however, were generally biologists or social scientists outside the mainstream of their fields. Within the mainstream of the social sciences, sociobiology was often greeted with general hostility and derision (Harris, 1979; Sahlins, 1976); it was criticized for being naïve and promoting an extreme reductionism (Freese, 1994).

Negative reaction to sociobiology among mainstream social scientists can be attributed to two broad points of view among social scientists. The first viewpoint is that sociobiology is so naïve and overly reductionistic that it ignores well-developed and predictive social and cultural theory developed within a legitimate scientific research strategy (Freese, 1994; Harris, 2001; Turner et al., 1997). The second viewpoint is associated with the post-modern, or hermeneutic, perspective that arose in the social sciences during the last half of the twentieth century. Opponents of this viewpoint reacted both to the biological reductionism and the rational, scientific program of attempting to develop an integrated, objective body of knowledge, or epistemology (Harris, 2001; Turner et al., 1997; Wilson, 1998). Post-modernists view an objective, scientific study of human social and cultural behavior as inappropriate. Instead, they argue that social and cultural studies should focus on a deeper understanding of the subjective, emic perspectives within a culture (Rorty, 1979).

The Biological Basis for Human Values Toward Wildlife: A Pleistocene Psychology?

We focus on ideas and research from sociobiology and evolutionary psychology (we treat these two labels as essentially synonymous throughout this chapter) and behavioral genetics in our discussion of a possible biological basis for human thought and behavior toward wildlife.

Sociobiologists and evolutionary psychologists propose that evolved neuro-biological traits shared by all humans influence complex human thoughts and behaviors (Barkow, Cosmides, & Tooby, 1992; Buss, 1999; Wilson, 1998). In contrast, behavioral geneticists primarily want to understand cognitive and behavioral similarities and differences among individuals by using quantitative genetics (family, adoption, and twin studies) and molecular genetics (studies of the effects of specific genes) (Plomin, DeFries, McClearn, & Rutter, 1997).

Sociobiologists and evolutionary psychologists propose that there is an evolved human psychology, shaped over millennia by the forces of natural selection and inclusive fitness on individuals, that forms the core of a universal human nature shared by all *Homo sapiens sapiens* regardless of cultural context (Buss, 1999; Cosmides & Tooby, 1992; Pinker, 2002; Wilson, 1998). Cosmides and Tooby (Cosmides & Tooby, 1992; Thornhill, Tooby, & Cosmides, 1997) argued that this *adapted mind* represents a complex structure of cognitive adaptations that address a large number of domain-specific decisions and behaviors.

In contrast to the notion that social and cultural contexts largely shape the psychology of individuals, evolutionary psychologists argue that psychology shaped human societies and cultures. This adapted psychology represents a complex structure of prepared and counter-prepared learning that enables humans to respond to diverse environmental contexts. A core assumption of this theoretical approach is that modern human psychology evolved as an adaptation for a life of hunting and gathering in a Pleistocene world during the 60,000–150,000 years of *Homo sapiens sapiens*. Thus, the modern human psychology we currently share developed before the rise of agriculture and large-scale permanent communities 5,000–10,000 years ago (Cosmides, Tooby, & Barkow, 1992). Further, evolutionary psychologists assume the past few thousand years is an insufficient amount of time to significantly alter the evolved psychology that was present at the inception of the species.

Three aspects of the sociobiology and evolutionary psychology literature are relevant to our examination of human–wildlife relations. All of these perspectives begin with the premise that the evolved human psychology was largely shaped by ecological forces selecting for adaptation to a hunting and gathering lifestyle. The first perspective argues that humans are "killer apes" prone to inter-specific and intra-specific aggression due to selection for hunting prowess beginning at least 1 million years before the present (YBP). The second perspective argues that humans have a deeply evolved psyche, the health of which is dependent upon close association with other animals. The third suggests that human relationships with wildlife are guided by the tendency for humans to anthropomorphize, a response that is borne from natural selection.

Killer Apes

While examining and describing the first australopithecine fossils, Dart argued the evidence suggested that these early hominids actively pursued and killed other large, mammalian species for food (Weiss & Mann, 1990). Since then, a variety of scholars have argued that hominoids are indeed killer apes and that an evolved hunting lifestyle helps explain the prevalence of violence in today's societies. The argument is that our evolved tendency to hunt and kill also made our species prone to intra-specific violence. In addition, this topic forms the basis of arguments made against contemporary recreational hunting, i.e., that hunting is a remnant of our savage past that we should somehow evolve past.

As the previous quote from Leopold (1922) attests, the idea that hunting is a deep, evolved trait among humans has existed for decades. Studies in paleoanthropology suggest that hunting, in at least some form, was present among the genus *Homo* at least 1 million YBP (Weiss & Mann, 1990) and continued through early archaic forms of *Homo sapiens* (200,000 to 60,000 YBP). Efficient hunting technologies and behaviors are well-accepted reasons for the rapid dominance of modern humans after their appearance approximately100,000 YBP.

Some scholars argued that selective pressure, in favor of hunting prowess and the ability to kill, led humans to be genetically predisposed toward aggression and killing (e.g., Ardrey, 1961; Weiss & Mann, 1990). In a now classic, yet contested work titled *Man the Hunter*, Washburn and Lancaster (1968, p. 299) provided the following perspective on the effects of hunting on aggressive tendencies:

> Men enjoy hunting and killing, and these activities continue in sports even when they are no longer economically necessary. If a behavior is important to the survival of a species (as hunting was for man throughout most of human history), then it must be both easily learned and pleasurable.

More recent proponents of this view include Wrangham and Peterson (1996). In *Demonic Males* they examined this issue by comparing human behavior to the behavior of our closest primate relative – the common chimpanzee (*Pan troglodytes*). They found that violent aggressive tendencies and a willingness to kill are traits humans shared with chimps. Wrangham and Peterson suggested that, while the violent and killer tendencies of human males are not necessarily caused by hunting, these killing tendencies make humans (and chimps) good hunters.

This is certainly a contested view. Megarry (1995) said, "It is not legitimate to propose that modern humans possess a 'biological basis for killing' which stands as an unreformed mental residue of a savage past" (p. 211). Megarry provided three basic reasons for this conclusion.

First, clear evidence of modern hunting is not found until approximately 40,000 YBP. For most of their history, humans and their predecessors (i.e., *Homo habilis, Homo ergaster, Homo erectus*) appear predominantly to have been scavengers and foragers. It was not until the appearance of archaic *Homo sapiens* that humans likely practiced the methods of hunting witnessed in modern hunter and gatherer societies, i.e., planned foraging patterns based on extensive environmental knowledge, sophisticated stone tools, advanced social organization, and representational art.

Second, most of the distinguishing characteristics of modern humans are not the products of pressures for an exclusive hunting lifestyle. These traits include bipedalism, manual dexterity, larger brains, and extensive material culture. Instead, these traits are more likely the products of a generalized foraging system.

Third, evidence in modern hunter and gatherer societies does not strongly support the idea that intra-specific aggression and violence evolved from a hunting lifestyle. Comparative societal studies suggest that while intra-human aggression and violence is universally present, aggression and violence appear to be heightened among food-producing societies as opposed to hunting and gathering societies. Most studies of hunters-gatherers do not find people of a highly aggressive nature; in fact, they tend to find calm, uncompetitive people (Megarry, 1995).

Hart and Sussman (2005) proposed an explanation on the role of hunting that has gained widespread acceptance over the view of man-the-hunter (Pickering, 2005). Their argument draws upon the archaeological and paleontology record and etiological comparisons between humans and near primates. They contended that the evolutionary roots of humans are not found in an existence centered on hunting; instead, it was shaped by a world in which humans were primarily a prey species, i.e., man-the-hunted. While there was a brief period in recent human history during which humans were true big game hunters, for millions of years, they were prey. Our role as prey had a profound effect on human development. "Ecologically and psychologically we were, until very recently, prey meat – meals for large frightening animals. It was a fact of life for our ancestors, and it is a fact of life for many humans today" (p. 247). The threat of predation, Hart and Sussman contended, shaped the hominid's body size, group living with multiple males, communication, bipedality, complex threat behaviors, and cognitive skills.

In summary, theories about the evolution of humans have focused on our interactions with wildlife, either in a role of hunter or as the hunted. Contemporary theories of evolution draw upon modern-day human fears of wildlife as evidence of evolutionary processes (Hart & Sussman, 2005). Assuming these explanations of how humans developed are accurate, it is easy to see that a significant component of human–wildlife relationships has an inherited basis.

Biophilia

In contrast to the thesis that a hunting or prey lifestyle leads to aggression or fear toward other species, some writers (e.g., Shepard, 1973; Wilson, 1984) have argued that an evolved nature to hunt has led to human appreciation of wildlife.

Building on findings from anthropology (Lee & Devore, 1968), Paul Shepard (1973) developed the thesis that the hunting and gathering lifestyle present among *Homo* spp. since the lower Pleistocene has been instrumental in shaping the human mind. As such, a desire to hunt or pursue animals is an innate characteristic of our species. In the *Tender Carnivore and the Sacred Game* (1973), Shepard also tied humans' evolved consumptive reliance on animals to an awe and reverence for those other species whose lives sustain ours.

In a similar vein, Wilson introduced the *biophilia hypothesis* (1984; 1993). Biophilia is "the innately emotional affiliation of human beings to other living organisms" (1993, p. 31). This tendency, Wilson argued, is inherited genetically and, similar to our ability for language, is part of our evolutionary history. Biophilia is facilitated by *prepared and counter-prepared learning* or biologically based tendencies to learn or resist learning.

While human behavioral theories derived from sociobiology and evolutionary psychology are challenging to test, advocates argue that Darwin's approach of *consilience of inductions* provides direction for testing hypotheses in evolutionary psychology (Ruse, 1989; Caporael, 2001). In this approach, multiple sources of evidence ranging from experiments, computer models, archival data, animal observation, and cultural comparisons provide a network for assessing the strength of hypothetical assertions. Caporael (2001) identified several evolutionary ideas that have been falsified through such approaches.

Though a relatively new area of study, a variety of authors present empirical findings assessing the validity of the biophilia hypothesis (Kellert & Wilson, 1993). For example, Ulrich (1993) found support for biophilia in research that examines biophobia. Biophobia deals with fears related to natural hazards or life forms such as snakes and spiders that have threatened humans throughout evolution. The research methods used in these studies are typically lab-based experiments that expose people to images (e.g., slides of snakes mixed with non-threatening photos). Sometimes these images are accompanied by brief electrical shock to produce a conditioned response. This allows the researchers to determine whether adverse response is learned more quickly and retained when associated with biophobic stimuli. The researchers measure change in physiological responses such as heart rate and skin conductance as images are shown. While reviewing this research, Ulrich (1993) concluded that defense responses to pre-modern, natural stimuli (like snakes) are easily learned and not easily forgotten.

Ulrich (1993) and Heerwagen and Orians (1993) found that research on landscapes tend to support the biophilia hypothesis. Research shows a pattern of positive response to certain natural settings – such as savanna-like environments and settings with water which, over millennia of early human existence, would be associated with primary necessities such as food, water, and security. Ulrich (1993) found that natural environments have a restorative effect, such as enhanced emotional states and physiological effects.

While definitive evidence confirming or disconfirming innate, affective-based human responses to nature is not forthcoming, one can think of many examples when human response to wildlife extends beyond cognitive reasoning. For example, an enduring irritation of wildlife and park managers is that park visitors feed wildlife. Repeated attempts to educate and regulate against this behavior have been frustrated by an unresponsive public.

During a follow-up interview investigation that we conducted at Rocky Mountain National Park in the early 1990s, we found that more than one-half of people who were observed to feed wildlife knew it was a violation of regulations for which they could be fined. The desire to be close simply outweighed the undesirable consequences. The notion of biophilia might lead us to the conclusion that wildlife feeding is the expression of an innate human tendency to benefit wildlife.

Katcher and Wilkins (1993) cautioned that biophilia couples an empirically testable hypothesis (i.e., humans have an innate tendency to focus on other animals) with a moral agenda (i.e., love and preservation of habitat and species). This coupling of testable ideas with a moral agenda makes objective assessments difficult because trying to falsify the hypothesis also may be interpreted as an attack on the moral argument. Katcher and Wilkins argued that the ethical connotations implied by the term *biophilia* need to be divorced from the notion that humans have an innate tendency to focus on other forms of life. They offered an explanation of the innate responses to wildlife that is similar to current views in developmental psychology. Human arousal may be an innate response, but the meaning is culturally derived. Katcher and Wilkins (1993) suggested that wildlife directs human focus (an innate response). More specifically, humans attend to the form or motion of living things and the way in which the form or motion patterns signal security or danger.

For example, a Heraclitean motion – aquarium fish swimming, waterfowl swimming, or animals grazing – suggests security while the erratic motions of injured or dying fish, birds taking flight, or animals breaking into a run suggest danger. Katcher and Wilkins (1993) found support for this concept in their work with patients that possess organic and functional mental disorders. Their work revealed that "the entry of animals to a purely human environment resulted in increased attention, increased social responding, positive emotion and, critically, speech" (p. 180).

Despite evidence that human response to animals is innate, Katcher and Wilkins do not support the notion that human response is innately *biophilic*. They noted that

> Children who throw stones at birds and children who feed birds are both responding to what may be an innate tendency to focus their attention on living things. The choice of the behavior used to engage the animal in the interaction is different and it is a *learned* behavior. (p. 175)

Thus, they proposed an interactionist perspective that integrates biologically innate tendencies and environmentally learned responses.

Kellert (1993) contended that the wide array of his "values toward wildlife" (e.g., utilitarian, naturalistic-ecologistic, scientific, aesthetic, etc.) are human expressions of biophilia. However, he offered no explanation of exactly how values are shaped by biology.

Is there a genetic tendency to be utilitarian or naturalistic? Or is there an innate tendency of humans to be aroused and attentive to wildlife, and that attention is given meaning by experience, culture, and learning? We believe it is the second explanation – specific human responses to wildlife represent an interaction between innate tendencies and culturally learned thoughts and behaviors. Humans appear to have innate tendencies to focus on and respond to natural objects, including wildlife. Such tendencies, however, cannot be portrayed as uniformly benevolent (i.e., biophilic) or malevolent in nature. Whether we love or eat animals, or both, in specific contexts represents an interaction of our evolved tendencies and learned thoughts and behaviors.

Wilson (1984) used the biophilia hypothesis to advocate a worldwide need for conservation and an environmental ethic. He suggested that the serious and growing loss of global biodiversity will produce a profound effect on the human psyche. We conclude ður review of biophilia with a cautionary quote from Katcher and Wilkins (1993), who questioned such an advocacy agenda:

> When we look at the source of love for other life-forms in our genetic inheritance we are searching our past for the authority to act on that love. That basis should cause us to question our own assumptions so that we restrain our human tendency to see things the way we want them to be. (p. 174)

Anthropomorphizing

Anthropomorphism refers to the human tendency to assign or infer human motives, intentions, and other characteristics to non-human animals. The tendency is ubiquitous in modern life and, according to Kennedy (1992), people anthropomorphize compulsively because the tendency is preprogrammed genetically. In post-industrial societies, anthropomorphizing is most obvious in entertainment media – such as books, movies, and cartoons – where non-human animals are portrayed as humans engaged in human-like social behavior. This tendency is not new; archaeological evidence suggests anthropomorphizing first emerged in the art of the Paleolithic period some 40,000 years ago (Mithen, 1996).

Rejection of the human tendency to anthropomorphize is integral to the Cartesian science view that animals are little more than machines (Kennedy, 1992). Strong norms against anthropomorphic tendencies are apparent among people engaged in the biological sciences and particularly the wildlife profession. It is argued that the tendency to project human characteristics on wildlife betrays the importance of objectivity and rational decision-making based on sound science. Kennedy, for example, contended that anthropomorphizing is a major problem in ethological investigations; he stated "...our penchant for anthropomorphic interpretations of animal behavior is a drag on the scientific study of the causal mechanisms of it" (1992, p. 5).

At the same time, however, the tendency to dismiss anthropomorphizing as an unacceptable characteristic for a person of science (similar to the trend regarding the study of emotion) has perhaps led social scientists to overlook its effect in directing human behavior and in understanding human–wildlife relationships.

Archaeologist Steven Mithen (1996) suggested that the increase of anthropomorphism resulted in a critical evolutionary improvement in human cognition. He contended that anthropomorphism, as well as enormous advances in culture, arose when humans attained the capability of integrating three separate areas of intelligence: natural history intelligence (e.g., interpreting natural symbols such as footprints), social intelligence (e.g., intentional communication), and technical intelligence (e.g., making artifacts from mental templates). While these intelligences remained separate in human thought through much of earlier human history, their integration marked a breakthrough that allowed the creation and use of visual symbols. The linkage is apparent in the creation of artifacts that not only served a utilitarian function but also had symbolic meaning and were used as a means of communication.

Mithen attributed the explosion of art in human society that occurred 40,000 years ago to this cognitive integration. Relevant to our interests, he found evidence for cognitive integration in the anthropomorphism reflected in this art. More specifically, drawings of figures intermixed with animal and human characteristics reflected the integration of social intelligence and natural history intelligence. It resulted from humans thinking of the natural world in social terms, with humans seen as animals and animals as humans.

Why did anthropomorphic thinking emerge as a human universal? Theoreticians propose anthropomorphic thinking emerged because it provided superior ability in hunting and gathering activities (Katcher & Wilkins, 1993; Kennedy, 1992; Mithen, 1996). As noted by Mithen, "Even though a deer or horse may not think about its foraging and mobility patterns in the same way as modern humans, imagining that it does can act as an excellent predictor for where the animal will feed and the direction in which it may move" (1996, p. 168).

Anthropomorphizing has been used by several researchers to explain aspects of post-industrial human relationships with animals. Katcher and Wilkins (1993) posited that the attention-focusing response that wildlife evokes from humans is due to the innate tendency to anthropomorphize. These researchers reported the positive effect that animals had on people in a clinical setting. In addition, Serpell (2003) proposed that pet-keeping can be explained because animals are seen as human-like and part of one's social network; this makes animals capable of giving social support and benefiting human health and well-being.

Manfredo and Teel in Chapter 8 suggest that anthropomorphic tendencies play a role, in combination with other factors, in facilitating shift in value orientations toward wildlife in North America. Modernization and material well-being have reduced the priority of values linked to existence needs. At the same time, the priority of belongingness and self-expression needs has been elevated. In this context, wildlife are increasingly viewed as companions instead of a food source, and that is, in part, facilitated by our anthropomorphism tendencies. They describe this as a transition from domination wildlife value orientations to mutualistic wildlife value orientations.

Behavioral Genetics: A Mechanism for Biological Influence on Human Thought and Behavior

While evolutionary psychology focuses on identifying evolved universal psychological characteristics, behavioral genetics focuses on identifying the origins of individual differences in human psychological characteristics. Although the evolved characteristics of a species clearly have genetic origins, differences among individuals within a species can and do arise for reasons other than genetics. The idea that differences in psychological characteristics among people are influenced by genetics extends back to the nineteenth century (Cosmides et al., 1997).

As in other areas of the social sciences, however, the idea that differences in psychological traits have genetic bases was used to justify racist political agendas. Misguided, and perhaps malevolent, synopses of research by Cyril Burt (1966), Arthur Jensen (1969), and more recently by Herrnstein and Murray (1994) argued that apparent differences in average intelligence between ethnic groups were genetically caused; this conclusion has cast suspicion on the field, and led to an intense criticism of behavioral genetic research. Most behavioral geneticists, however, view such group-level comparisons as a fundamentally flawed approach (Plomin et al., 1997), and propose a focus on individual-level effects. These researchers find a wealth of inferentially strong evidence indicating that genetics play an influential role on everything from cognitive abilities and disabilities to personality and attitudes (Plomin et al., 1997). Lykken (2006) said, "It is by now generally accepted that most consistent individual differences – psychological, physiological or psychophysiological – are from partly to strongly heritable" (p. 306).

Psychopathology and Personality

Genetic research on psychopathology (mental illness) has been extensive since 1970 (Plomin et al., 1997). Family, twin, and adoption studies have found strong genetic influence on some severe disorders such as schizophrenia and bipolar mood disorder, while evidence suggests only a weak genetic influence on general depression. Comparatively little research has been conducted on anxiety and fear disorders and the results are quite mixed depending on the specific disorders; however, both panic disorder and generalized anxiety disorder appear to have a substantive heritability. Little research has been conducted on specific phobias, but existing research suggests a modest genetic influence (Kendler, Neale, Kessler, Heath, & Eaves, 1992; Rose & Ditto, 1983). Disorders of childhood such as autism (Rutter, Bailey, Bolton, & LeCouteur, 1993; Bailey, Philips, & Rutter, 1996) and hyperactivity (Silberg et al., 1996) demonstrate a strong genetic influence; however, conduct disorders without a hyperactivity component show almost no genetic influence (Silberg et al., 1996). The strength of evidence concerning the genetic influence on psychopathology is so compelling that recent arguments in the literature consider both genetic and environmental causal effects when explaining psychopathological behavior (Rutter, Pickles, Murray, & Eaves, 2001).

Although somewhat less robust, there is also a large body of research which supports a strong genetic influence ($h \sim 0.30$–0.50) on personality traits and disorders (Bouchard, Lykken, McGue, Segal, & Tellegen, 1990; Bouchard & Loehlin, 2001; Loehlin, 1992; Plomin et al., 1997).

Bouchard and Loehlin (2001) provided a thorough review of recent twin, adoption, and family research studies that used a variety of personality-measurement approaches and tools. Twin studies suggest a broad heritability of about 0.50 for most personality traits, while adoption and family studies suggest a more modest 0.30. Depending on the measurement scale used, some of these personality traits (e.g., social responsibility, traditionalism) appear similar to what other researchers consider values. For this reason, Bouchard and Loehlin (2001) urged more research to examine the behavioral genetics of values and attitudes.

Heritability of Values and Attitudes

Despite a large body of research indicating substantial heritability of psycho-pathologies and personality traits, as recently as 1990 Tesser and Shaffer reviewed the attitude research and identified no scholarly articles on the heritability of attitudes. Likewise, the most thorough review of attitudes used in academic instruction (Eagly & Chaiken, 1993) paid scant attention to biological or genetic influences on attitudes. As reported by Tesser (1993), McGuire's statements that "even theorists who agree on little else are in complete accord on the extreme and undemonstrated notion that all attitudes are developed through experience" (1969, p. 161) and "attitude theorists typically abhor hypothesizing genetic influence" (1985, p. 253) remain largely accurate. Such perspectives continue to dominate the field despite findings from behavioral genetics that broad attitudes such as author-itarianism (Scarr & Weinberg, 1981) as well as job satisfaction (Arvey, Bouchard, Segal, & Abraham, 1989), work values (Keller, Bouchard, Arvey, Segal, & Davis, 1992), and religious values (Waller, Kojetin, Bouchard, Lykken, & Tellegen, 1990) all have heritability ranging from 0.20 to 0.50.

In the past decade, however, mainstream attitude researchers explore the relevance of heritability for attitudes (Crelia & Tesser, 1996; Olson, Vernon, Harris, & Jang, 2001; Tesser, 1993). Building on the work of Eaves, Eysenck, and Martin (1989), Tesser (1993) suggested that specific attitudes vary in their level of heritability. Some attitudes (e.g., severe criminal punishment and religious attitudes) have substantial levels of heritability (0.40–0.60), while others (e.g., attitudes toward extramarital sex) have no heritability. Recently, Olson et al. (2001) provided more evidence of differential heritability across specific attitudes and found that attitudes that were more heritable are also psychologically stronger.

Lykken, Bouchard, McGue, and Tellegen (1993) offered compelling evidence that there is a genetic basis that guides our interests and our leisure pursuits. Data for this study were obtained from a survey that employed 100 items measuring occupational interests and 120 items measuring leisure interests. Subjects were obtained from the Minnesota twin registry that included pairs of twins born in Minnesota from 1936 to 1955. Nine hundred twenty-four twin pairs raised together and 94 twin pairs raised apart were included in the study. By comparing the correlations of the monozygotic twins (identical genes) and dizygotic (half of the genes are identical) it is possible to explore what types of traits are inherited as well as the way in which the transmission process occurs (whether it appears to rest on a single or a combination of genetic information). By separating those that live apart and together, it is possible to partition out effects due to upbringing. Heritability estimates are established using a statistic that ranges from 0, no correspondence of participation between twins in the sample, to 1, where each pair would respond identically across all subjects in the sample. Results of the study showed that mean heritability estimates were 0.57 for individual items and 0.65 for clusters of items measuring leisure and occupational interest items. This suggests a very strong influence of genetics on one's leisure and occupational pursuits.

It would be easy to misrepresent this effect. A person is not genetically predisposed to engage, for example, in team sports. Lykken et al. (1993) are careful to say that interests are learned traits. The authors contended, however, that traits such as physique, aptitude, temperament, and personality are the focal point of genetic influence and that these characteristics, when combined with a cafeteria of experience, direct a person's choice; (i.e., a person would not select something that is not available or that he has not heard of).

Attitudes Toward Wildlife and Wildlife Values

No one has conducted research explicitly on the heritability of wildlife values, and essentially no research on this topic was available for review by authors contributing chapters to *The Biophilia Hypothesis* (1993). However, the item inventory used by Lykken et al. (1993) included such items as

- "Hunting small game"
- "Hunting big game"
- "Going fishing"
- "Going on camera safaris in African, Borneo, or the Amazon Basin"
- "Wildlife study, bird watching"
- "Canoe trips in wilderness areas"
- "Working with animals, training or showing dogs, horses, cats, etc."

Respondents also completed inventories addressing occupational interests (100 occupations including veterinarian, naturalist, wildlife biologist, and game warden), self-rated talents (including working with animals), and a self-rating of personality characteristics (including *empathy for animals* defined as a sympathy and concern for animals, an interest in and affection for most living things, an ability to enjoy the company of animals, and a concern for their welfare).

Among the 39 factors that were identified from the list of interests, three relate to wildlife, nature or animals: blood sports (i.e., hunting and fishing, 6 items, $a = 0.87$), wilderness activities (4 items, $a = 0.74$), and working with animals (11 items, $a = 0.86$) (Lykken et al., 1993). For each of these dimensions, heritability ranged between 0.40 and 0.55 for both males and females. In addition, when looking at hunting and fishing interests for twins reared apart, the dizygotic twins had a correlation of zero compared to 0.43 for monozygotic twins (Lykken, personal communication).

This finding suggests that interest in hunting or fishing is what Lykken refered to as "emergenic" (2006). This means that the observed trait is the result of several genes contributing to create the effect. Emergenic traits do not necessarily run in families (as indicated by the lack of correlations among non-identical twins) (Lykken, McGue, Tellegen, & Bouchard, 1992). Other traits identified as emergenic are: neuroticism, aggressiveness, impulsiveness, positive emotionality, artistic interests, and vocal quality (Lykken, 2006).

What can we conclude from the existing research on the heritability of attitudes and interests? As Olson et al. noted (2001), "Attitudes are learned. But attitudes also depend on biological factors" (p. 859). Research by Lykken et al. (1993) suggested this is true in the specific case of attitudes toward and interest in wildlife use, nature-based recreation, and working closely with animals. Because such attitudes appear to be strongly inheritable, they may be held much more strongly and be more resistant to change within the individual (Olson et al., 2001). These data do not imply anything about the notion of a universal biophilia present in humans.

Conclusion

Within the past 30 years, perspectives on the importance of biological inheritance have changed greatly. Once suggestive of repugnant political implications, current views suggest that inheritance shapes predispositions that affect how we perceive, learn, and respond to our environment. In this chapter, we explored questions regarding the inherited tendencies of humans that would affect human relationships with wildlife. Among the many topics that loomed in the cognitions of humankind, none have been as enduring as wildlife. As an enduring source of human fear, security, and existence, our relationships with wildlife are a key place to look for traces of inherited tendencies.

From the view of evolutionary psychology, we explored the notion that there are certain human universals that dictate response to wildlife. We explored three different topics: aggression, biophilia, and anthropomorphizing.

From the view of behavioral genetics we explored the question of whether individual values and attitudes concerning wildlife and the human use of wildlife are due to factors of heritability. A large body of evidence suggests genetics influence many psychological traits and cognitions, including individual interests in using, protecting, and interacting with wildlife and nature. The literature does not suggest that such psychological tendencies are not totally genetically predetermined, and individual differences in such characteristics are clearly influenced by an interaction of biology and experience (e.g., learning). More research on the role of biological factors, including genetics, in forming values and attitudes and change in those values and attitudes will help us understand their formation and malleability.

Key Points Regarding the Biological Basis of Human–Wildlife Interactions

- A significant portion of modern-day human responses to wildlife may be genetically prepared, more than might have been thought just two decades ago.
- Researchers debate whether human universals are shaped by natural selection. The chapter explores three topics that appear particularly relevant to understanding human–wildlife relationships – aggression and hunting, biophilia, and anthropomorphism.
- There are countervailing thoughts on how our role in the predator–prey relationships of early hominid existence might have shaped our current humanity. One view, expressed as *man-the-hunter*, proposes that for millions of years of evolution, humans were hunters and natural selection favored the aggressive tendencies in humans that would make them good hunters. In contrast, contemporary approaches propose the view of *man-the-hunted*. Our role as prey species affected our social organization and our psychology. This shapes our caution and fear responses, particularly to large predators.
- Biophilia suggests there is an innate human dependency on, and positive affective response to, natural environments. Some evidence suggests there may be an innate biophobic (fear arousal) response to threats in the natural environment and positive affective response to some types of natural environmental features. Some care must be taken in discussions of biophilia because it appears to merge a political advocacy position with issues of scientific interest.
- Anthropomorphism is the universal tendency of humans to attribute human characteristics to wildlife (and other non-human entities). It appears to be a biologically prepared response among humans that arose about 40,000 years ago when humans attained *integrative* cognitive processing. It is hypothesized to have emerged because it would have improved human abilities as hunters. The tendency to anthropomorphize may play an important role, along with other factors in a changing society, in how humans interact and respond to wildlife.

- In addition to the search for a genetic explanation of human universals, research has also attempted to explain the variability of human behavior due to genetic influences. For a good portion of the twentieth century, psychology ignored this area of inquiry due to political implications associated with supporting bigotry and discrimination. However, recent research is rapidly amassing evidence of a strong genetic effect on a wide range of human traits. Twin studies provide compelling evidence that show genetic influence, combined with one's environmental surroundings, has a strong influence on human choice.

Management Implications

While information about the biological basis of human–wildlife relationships may have little to offer in directing day-to-day management decisions, it has much to offer our philosophical orientation toward human–wildlife interactions. An enduring question asks how much of our behavior and interest is shaped culturally. The mid-twentieth-century view was that heritability had little to do with human behavior. But recent views challenge that assumption. Findings would suggest that there are (a) human universals regarding response to wildlife and (b) there are individual differences among people that can be explained by heritability.

It would appear that, at some level, human attraction and repulsion to wildlife has a strong biological basis. There are three areas of ongoing concern to management that might benefit from a consideration of this. First, humans appear to be naturally drawn to wildlife and their close presence appears to have a therapeutic or restorative effect on people in some situations. Can managers create appropriate experiences for people that enhance this benefit to people? Second, persuasive communication might be more effective if it attends to the basic elements of attraction and repulsion (this is also examined in Chapter 3). Third, situations where managerial cognitive appeals about wildlife are understood yet ignored (e.g., wildlife feeding) might signal the operation of these more basic biologically prepared causes. If this is the case, we should consider alternatives (structural or design strategies) to cognitive reasoning in dealing with such problems (when possible).

While individual differences suggest a strong biological basis in choices, such choices are culturally determined through the structuring of alternatives available to people. More specifically, there appears to be no basis in the argument that recreational hunting or fishing is the enactment of a biologically prepared tendency. Heritability might affect the traits of an individual that predispose her to engage in that *type* of activity; however, the specific type of activity (e.g., hunting, fishing, hiking) cannot be determined because there are many alternatives to those specific forms of engagement, and not all of these activities may be available to the person.

References

Ardrey, R. (1961). *African genesis*. New York: Atheneum.

Ardrey, R. (1966). *The territorial imperative*. New York: Atheneum.

Arvey, R. D., Bouchard, T. J., Segal, N. L., & Abraham, L. M. (1989). Job satisfaction: Environmental and genetic components. *Journal of Applied Psychology, 74*, 187–192.

Bailey, A., Philips, W., & Rutter, M. (1996). Autism: towards an integration of clinical, genetic, neuropsychological and neurobiological perspectives. *Journal of Child Psychology and Psychiatry, 37*, 89–126.

Barkow, J. H., Cosmides, L., & Tooby, J. (1992). *The adapted mind: evolutionary psychology and the generation of culture*. New York: Oxford University Press.

Bouchard, T. J., Lykken, D., McGue, M., Segal, N. L., & Tellegen, A. (1990). Sources of human psychological differences: The Minnesota study of twins reared apart. *Science, 250*, 223–228.

Bouchard, T. J., & Loehlin, J. C. (2001). Genes, evolution and personality. *Behavior Genetics, 31*, 243–273.

Burt, C. (1966). The genetic determination of differences in intelligence: a study of monozygotic twins reared together and apart. *British Journal of Psychology, 57*, 137–153.

Buss, D. (1999). Evolutionary psychology: The new science of the mind. Boston: Allyn Bacon.

Caporael, L. R. (2001). Evolutionary psychology: Toward a unifying theory and hybrid science. Annual Review of Psychology, 52,607–628.

Clossen, C. C. (1897). The hierarchy of European races. *American Journal of Sociology, 3*, 314–327.

Cosmides, L., & Tooby, J. (1992). Cognitive adaptations for social exchange. In J. Barkow, L. Cosmides, & J. Tooby (Eds.), *The adapted mind*, New York: Oxford University Press.

Cosmides, L., Tooby, J. H., Turner, J. H., & Velichkovsky, B. M. (1997). Biology and psychology. In P. Weingart, S. D. Mitchell, P. S. Richerson, & S. Maasen (Eds.). *Human by nature: between biology and the social sciences*. Manhwah, NJ: Erlbaum Associates.

Cosmides, L., Tooby, J, & Barkow, J. H. (1992). Introduction: evolutionary psychology and conceptual integration. In J. H. Barkow, L. Cosmides, & J. Tooby (Eds.), *The adapted mind: evolutionary psychology and the generation of culture*. New York: Oxford University Press.

Crelia, R. A., & Tesser, A. (1996). Attitude heritability and attitude reinforcement: a replication. *Personality and Individual Differences, 21*, 803–808.

Dawkins, R. (1976). *The selfish gene*. Oxford: Oxford University Press.

Eagly, A. H., & Chaiken, S. (1993). *The psychology of attitudes*. San Diego, CA: Harcourt, Brace, & Jovanovich.

Eaves, L. J., & Eysenck, H. J., & Martin, N. G. (1989). *Genes, culture and personality: An empirical approach*. San Diego, CA: Academic Press.

Freese, L. (1994). The song of sociobiology. *Sociological Perspectives, 37*, 337–374.

Harris, M. (1999). *Theories of culture in postmodern times*. Walnut Creek, CA: Altamira Press.

Harris, M. (2001). Cultural materialism: the struggle for a science of culture (updated edition). Walnut Creek, CA: Altamira Press.

Hart, D., & Sussman, R. W. (2005). *Man the hunted.: Primates, predators, and human evolution*. New York: Westview.

Hawley, A. H. (1986). Human ecology: a theoretical essay. Chicago: The University of Chicago Press.

Heerwagen, J. H. & Orians, G. (1993). Humans, habitats and aesthetics In S. R. Kellert, & E. O. Wilson (Eds.), *The biophilia hypothesis (pp. 138–172)*. Washington DC: Island Press.

Herrnstein, R. J., & Murray, C. (1994). *The bell curve: intelligence and class structure in American life*. New York: Free Press.

Inglehart, R. (1990). *Culture shift in advanced industrial societies*. New Jersey: Princeton University Press.

Inglehart, R., & Baker, W. E. (2000). Modernization, cultural change, and the persistence of traditional values. *American Sociological Review, 65*, 19–51.

Jensen, A. R. (1969). Intelligence, learning ability and socioeconomic status. *Journal of Special Education, 3*, 23–35.

Katcher, A., & Wilkins, G. (1993). Dialogue with animals: Its nature and culture. In S. R. Kellert, & E. O. Wilson (Eds.), *The biophilia hypothesis* (pp. 173–200). Washington DC: Island Press.

Keller, L. M., Bouchard, T. J., Arvey, R. D., Segal, N. L. & Davis, R. V. (1992). Work values: Genetic and environmental influences. *Journal of Applied Psychology, 77*, 79–88.

Kellert, S. R. (1993). The biological basis for human values of nature. In S. R. Kellert, & E. O. Wilson (Eds.), *The biophilia hypothesis* (pp. 42–69). Washington, DC: Island Press.

Kellert, S. R., & Wilson, E. O. (1993). *The biophilia hypothesis*. Washington, DC: Island Press.

Kendler, K. S., Neale, M. C., Kessler, R. C., Heath, A. C., & Eaves, L. J. (1992). A population-based twin study of major depression in women: The impact of varying definitions of illness. *Archives of General Psychiatry, 49*, 257–266.

Kennedy, J. S. (1992). *New anthropomorphism*. Cambridge: University Press.

Lee, R. B., & Devore, I. (1968). *Man the hunter*. Chicago: Aldine Publishing.

Leopold, A. (1922). Goose music. In Leopold. L. B. (Ed.), *Round river: from the journals of Aldo Leopold*. New York: Oxford University Press.

Loehlin, J. C. (1992). *Genes and environment in personality development*. Newbury Park, CA: Sage.

Lykken, D. T. (2006). The mechanism of emergenesis. *Genes, Brain and Behavior, 5*, 306–310.

Lykken, D. T., Bouchard, T. J., McGue, M., & Tellegen, A. (1993). Heritability of interests: a twin study. *Journal of Applied Psychology, 78*, 649–661.

Lykken, D. T., McGue, M., Tellegen, A., & Bouchard, T. J. (1992). Emergenesis: Genetic traits that may not run in families. *American Psychologist, 47*, 1565–1577.

Lykken, D (personal communication, date of communication needed).

Mazur, A., & Robertson, L. S. (1972). *Biology and social behavior*. New York: Free Press.

Megarry, T. (1995). *Society in prehistory*. Washington Square: New York University Press.

McGee, R. J., & Warms, R. L. (1996). *Anthropological theory: an introductory history*. Mountain View, CA: Mayfield Publishing Co.

McGuire, W. J. (1969). The nature of attitudes and attitude change. In G. Lindzey, & E. Aronson (Eds.), *The handbook of social psychology* (2 nd ed., pp. 136–314). Reading, MA: Addison-Wesley.

McGuire, W. J. (1985). Attitudes and attitude change. In G. Lindzey, & E. Aronson (Eds.), *The handbook of social psychology* (Vol. 2, pp. 233–346). New York: Random House.

Mithen, S. (1996). *The prehistory of the mind: The cognitive origins of art, religion and science*. New York: Thames and Hudson.

Morris, D. (1967). *The naked ape*. New York: McGraw-Hill.

Olson, J. M., Vernon, P. A., Harris, J. A., & Jang, K. L. (2001). Heritability of attitudes: a study of twins. *Journal of Personality and Social Psychology, 80*, 845–860.

Opler, M. E. (1945). The bio-social basis of thought in the Third Reich. *American Sociological Review, 10*, 776–785.

Pinker, S. (2002). *The blank slate: the modern denial of human nature*. New York: Viking Press.

Plomin, R., DeFries, J. C., McClearn, G. E., & Rutter, M. (1997). *Behavioral genetics* (3rd ed.). New York: W.H. Freeman and Company.

Quammen, D. (2003). *Monster of god: The man-eating predator in the jungles of history and the mind*. New York: W.W. Norton.

Rose, R. J., & Ditto, W. B. (1983). A developmental-genetic analysis of common fears from early adolescence to early adulthood. *Child Development, 54*, 361–368.

Rorty, R. (1979). *Philosophy and the mirror of nature*. New Jersey: Princeton University Press.

Ruse, M. (1989). *The Darwinian paradigm*. New York: Routledge.

Rutter, M., Bailey, A., Bolton, P., & LeCouteur, A. (1993). Autism: Syndrome definition and possible genetic mechanisms. In R. Plomin, & G. E. McClearn (Eds.), *Nature, nurture and psychology* (pp. 269–284). Washington, DC: American Psychological Association.

Rutter, M., Pickles, A., Murray, R., & Eaves, L. (2001). Testing hypotheses on specific environmental causal effects on behavior. *Psychological Bulletin, 127,* 291–324.

Sahlins, M. (1976). *The use and abuse of biology: An anthropological critique of sociobiology.* Ann Arbor: University of Michigan Press.

Saurin, J. (1993). Global environmental degradation, modernity, and environmental knowledge. *Environmental Politics, 2*(4), 46–64.

Scarr, S., & Weinberg, R. A. (1981). The transmission of authoritarianism in families: Genetic resemblance in social-political attitudes? In S. Scarr (Ed.), *Race, social class, and individual differences in IQ.* Hillsdale, NJ: Erlbaum.

Schwabe, C. (1994). Animals in the ancient world. In A. Manning, & J. Serpell (Eds.), *Animals and human society* (pp. 36–58). London: Routledge.

Serpell, J. A. (2003). Anthropomorphism and anthropomorphic selection – beyond the "cute response." *Society and Animals, 11*(1), 83–101.

Silberg, J. L., Rutter, M. L., Meyer, J., Maes, H., Hewitt, J. K., Simonoff, E., et al. (1996). Genetic and environmental influences on the covariation between hyperactivity and conduct disturbance in juvenile twins. *Journal of Child Psychology and Psychiatry, 37,* 803–816.

Skinner, B. F. (1971). *Beyond freedom and dignity.* New York: Bantam.

Shepard, P. (1973). *The tender carnivore and the sacred game.* Athens: University of Georgia Press.

Smith, E. A., & Wishnie, M. (2000). Conservation in small-scale societies. *Annual Review of Anthropology, 29,* 493–524.

Steward, J. (1977). *Evolution and ecology.* Urbana: University of Illinois Press.

Steward, J. H., Steward, J. C., & Murphy, R. F. (1977). *Evolution and ecology: essays on social transformation.* Urbana: University of Illinois Press.

Tesser, A. (1993). The importance of heritability in psychological research: the case of attitudes. *Psychological Review, 100,* 129–142.

Tesser, A., & Shaffer, D. (1990). Attitudes and attitude change. In M. R. Rosenzweig, & L. Porter (Eds.), *Annual review of psychology* (Vol. 41, pp. 479–523). Palo Alto, CA: Annual Reviews.

Thornhill, N., Tooby, J., and Cosmides, L. (1997). Introduction to evolutionary psychology. In P. Weingart, S. D. Mitchell, P. J. Richerson, & S. Maasen (Eds.), *Human by nature: Between biology and the social sciences* (pp. 212–238). Mahwah, NJ: Lawrence Erlbaum.

Tiger, L. (1969). *Men in groups.* New York: Random House.

Tooby, J., & Cosmides, L. (1992). The psychological foundations of culture. In J. H. Barkow, L. Cosmides, & J. Tooby (Eds.), *The adapted mind: evolutionary psychology and the generation of culture* (pp. 19–136). New York: Oxford University Press.

Turner, J. H., Mulder, M. B., Cosmides, L., Giesen, B., Hodgson, G., Maryanski, A. M., et al. (1997). Looking back: historical and theoretical context of present practice. In P. Weingard, S. D. Mitchell, P. J. Richerson, & S. Maasen (Eds.), *Human by nature: between biology and the social sciences* (pp. 17–65). Mahwah, NJ: Lawrence Erlbaum.

Ulrich, R. S. (1993). Biophilia, biophobia, and natural landscapes. In S. R. Kellert, & E. O. Wilson (Eds.), *The biophillia hypothesis* (pp. 73–137). Washington, DC: Island Press.

Washburn, S. L., & Lancaster, C. S. (1968). The evolution of hunting. In R. B. Lee, & I. Devore (Eds.), *Man the Hunter* (pp. 293–303). Chicago: Aldine Publishing.

Weiss, M. L., & Mann, A. E. (1990). *Human biology and behavior: an anthropological perspective* (5th ed.). Boston: Little, Brown and Co.

Waller, N. G., Kojetin, B. A., Bouchard, T. J. Jr., Lykken, D. T., & Tellegen, A. (1990). Genetic and environmental influences on religion interests, attitudes and values: A study of twins reared apart and together. *Psychological Science, 1,* 138–142.

Wilson, E. O. (1975). *Sociobiology: The new synthesis.* Cambridge: Harvard University Press.

Wilson, E. O. (1984). *Biophilia: The human bond with other species*. Cambridge: Harvard University Press.

Wilson, E. O. (1993). Biophilia and the conservation ethic. In S. R. Kellert, & E. O. Wilson (Eds.), *The biophilia hypothesis* (pp. 31–41). Washington, DC: Island Press.

Wilson, E. O. (1998). *Consilience: the unity of knowledge*. New York: Vintage.

Wrangham, R., & Peterson, D. (1996). *Demonic Males: Apes and the origins of human violence*. Boston: Houghton Mifflin.

Chapter 3
Understanding the Feeling Component of Human–Wildlife Interactions

Contents

Introduction

I was raised in Kane, Pennsylvania, a timber town surrounded by the Allegheny National Forest. Our house was located on the southern edge of town with forest bordering two sides. One lazy summer evening when I was 8 years old my best friend, his younger brother, and I decided to walk to the junior high school basketball courts. From our house, we walked up the hill and took a left-hand turn toward the school, which was just a block away. While walking toward the school, we had a unique vantage point. We could see the length of an open field that surrounded the back side of the school and opened from the patch of trees behind my home. As we wandered along the sidewalk, my friend's younger brother opened his mouth and eyes wide, stammered, and pointed toward the field. I spun to see a large black bear ambling across the field! The bear was no more than 75 yards away but seemed oblivious to our presence.

M.J. Manfredo, *Who Cares About Wildlife?*,
DOI: 10.1007/978-0-387-77040-6_3, © Springer Science+Business Media, LLC 2008

As I know now, my amygdale, in the temporal lobe of my brain, received this horrifying information and instantly sent signals to the rest of my body to prepare me for flight. My heart pounded, my muscles contracted, and I began to sweat. I focused on just one thing: running as fast as I could back to my home. I can still remember the sensation of running downhill, being on the edge of bodily control and thinking that at any second I might fall forward and tumble. But I was motivated by the younger brother who was yelling that the bear was chasing us and I, for one, did not want to be last in line.

Breathless and frantic we arrived back at the house, and, until our parents began to question us, we did not notice that the younger brother had wet his pants. Interestingly, my recollections of childhood are few, yet this memory of seeing the bear is extremely vivid, as though it happened in my recent adult life.

Emotion researchers would conclude that this experience is highly memorable because it was stored with the strong sense of fear. Emotions are the hot feelings that mix with the cold realm of rational thought and give felt meaning to life. I would invite you, the reader, to stop and think of your most memorable experiences that involve wildlife. When recalling events, think about the responses you had that made this experience so memorable to you. Do terms such as happiness, surprise, sadness, fear, or anger come to mind? I suspect the answer is "yes," as it is those responses that reveal the personal significance of an event.

As central as emotions may be in our personal experiences with wildlife, there is very little research on this topic directly. The preponderance of research in the area of human dimensions of wildlife (HDW) comes from the cognitive tradition of social psychology. One of the most significant reasons for this neglect of emotions can be attributed to a cultural belief that emotions interfere with more desirable forms of human response. As noted by Cacioppo and Gardner (1999),

> An assumption by rationalists dating back to the ancient Greeks has been that higher forms of human existence – mentation, rationality, foresight, and decision-making – can be hijacked by the pirates of emotion. In accordance with the classic assumption that emotion wreaks havoc on human rationality, the emphasis for years in psychology has been on cognition and rationality. . . . (p. 194)

In the context of that rationality, it has been hard to justify the applicability of HDW research findings on people's emotions. In fact, the ideal of the wildlife professional has been to emphasize *science* while striving to exclude emotional considerations from the decision-making process. Those who have been involved with wildlife decision-makers have most certainly heard someone dismiss a stakeholder's view as being "just emotional." That view has done little to encourage HDW researchers to embark upon an exploration of the role of emotion in human–wildlife relationships.

Another reason for the lack of attention given to emotional responses in HDW research is attributed to methodological challenges. A good deal of research in the area of emotions often employs techniques that use physiological measures which necessitate laboratory-based, experimentally designed studies.

Moreover, findings from studies using these types of methods may have low external validity and limited implication for an applied field such as HDW.

Even with these concerns, the exploration of emotional response to wildlife may be one of the most intriguing and fruitful areas for future investigation. First, emotions may be the most basic human response to animals. As the brief review of literature below indicates, emotions are believed to be inherited human responses that emerged to provide evolutionary advantage. Wildlife were prominent in early hominid life – either as prey or as threats to survival – and it would be hard to deny that, via emotions, humans have inherited responses to wildlife. That is not to suggest that cognitions and cultural learning do not override or rechannel these emotional responses in most cases. However, it would appear that rudiments of emotion are inherited and interact closely with cognitive functions to affect human behavior.

Another reason for increasing the attention given to emotions is the growing role they play in western society. Fischer, Manstead, Evers, Timmers, and Valk (2004) contend that the shift toward increasing emotionality in social life is part of a larger cultural development in the late twentieth century. Increasingly, emotions are displayed in social settings and are expected to be authentic. The desire to reveal emotions is thus an increasingly critical component of communication. An important question asks how to incorporate emotional communication into discourse about natural resources (Milton, 2002).

Along that line, research increasingly shows that, while emotions may produce uncontrolled behavior (e.g., rage), they also play a critical role in sound decision making and are an important component of intelligence (Cacioppo and Gardner, 1999). Improvement in understanding human behavior will ultimately be obtained by understanding the interrelationship of cognitive concepts such as attitudes, values, and norms with affective concepts such as mood and emotion.

Finally, from an applied perspective, it is important to realize that emotional responses are at the heart of human attraction to, and conflict over, wildlife. We need a better understanding of the operation of emotions and the effect that emotions can have on our behaviors and social interactions. This holds hope for improving techniques used to communicate with stakeholders, understanding their involvement and attachment to natural resources, and achieving conflict resolution and consensus building.

This chapter reviews several fundamental concepts about emotions and their relationship to cognitions.

Emotions Are Part of Affect

The term *affect* is used to refer to the general class of *feeling states* experienced by humans. The topic of mood and emotions is subsumed under this category (Rosenberg, 1998). Emotions are different than moods because emotions are about a specific event, have short duration, and occupy conscious thought.

Alternatively, moods are conceptualized as longer-lasting affective conditions and in the background of consciousness (Fredrickson & Branigan, 2001). Clearly, emotions and mood are intertwined, and, as noted by Fredrickson and Branigan (2001), the distinction may be more theoretical than practical. In this chapter, I focus primarily on emotions, but, where appropriate, I introduce occasional findings that deal with affect generally or with mood.

Types of Emotions

Is there a parsimonious way to categorize the various emotions that people might experience? Many researchers have proposed typologies of emotions, and, to date, none stands alone as the single authority on the topic. However, many researchers argue that emotions should be divided into those that are primary (or basic) and those that are secondary. In a review of typologies of emotions, Kemper (1987) suggested that basic (or primary) emotions include those that (a) can be observed or inferred in most animals, (b) are found in all cultures, (c) appear early in development of humans, and (d) have distinct patterns of autonomic physiological response. He proposed a short list of basic emotions that included fear, anger, sadness, and satisfaction.

Two other commonly cited typologies are Izard's (1977) list of fear, anger, enjoyment, interest, disgust, joy, surprise, shame, contempt, distress, and guilt and Eckman's (1984) fear, anger, sadness, happiness, disgust, and surprise. More recently, Barrett, Mesquita, Ochsner, and Gross (2007) advocated use of the concept of core affect in studying emotions. This proposes a simple affective dichotomy of pleasure and displeasure to which additional meaning is overlaid based on past knowledge as well as situational and contextual factors.

Secondary emotions are defined as blends of primary emotions. Secondary emotions are illustrated by Plutchik (2003), who conducted a study to determine which primary emotions constitute secondary ones. He asked a group of study participants to judge what combinations of basic emotions make secondary ones. Combinations on which there was high agreement suggested that remorse is a mixture of sadness and disgust, that hatred combined disgust and anger, that pride included anger and joy, that shame mixed fear and disgust, and that anxiety was comprised of anticipation and fear.

Conceptual Approaches to Emotions

As with other concepts addressed in this book, the concept of emotion has many different definitions. Cornelius (1996) summarized five separate components that emerge from the many approaches to this complex topic. Early attention given to the topic of emotions examined the consistency of *expressive reactions*, which result from an emotion-arousing event (smiling, frowning, showing surprise), and the *physiological reactions* of the body (increased heart rate, skin temperature, blood flow). For some theorists, these are the defining elements of emotion.

Another critical characteristic of emotion is the nature of behavioral *coping* that might occur as a result of an event and emotional reactions. This would include behaviors such as fleeing, approaching, fighting, or spitting.

Another major area of response deals with the thoughts or *cognitions* that arise at various points in the emotion-eliciting process. For example, interpreting the cause of a situation would distinguish between widely different emotions such as sadness and joy. Finally, there is the element of *feelings* which are the bodily sensations and *subjective experiences* people have. The element of feelings is one we most readily identify in our day-to-day use of language about emotions. When people talk about their joy and happiness over an event or the anger or fear they felt, we tend to identify with the personal sensation of that experience.

While the experiential element of emotions has been neglected in the past, researchers have recently recognized the importance of attaining a better explanation of this component of emotion (Barrett, Mesquita, Ochsner, & Gross, 2007).

As noted by Cornelius (1996), definitions of emotions vary based on which of these elements is the focus of one's theory. The following descriptions, adapted from Cornellius, represent the prominent approaches to emotion theory.

Darwin and the Importance of Emotional Display

In the mid-1800s, Charles Darwin advanced a theory of evolution that suggested that types of animals, their shape, and anatomy were an adaptive response to changing environments. Gradual change within species occurred through the forces of selection, where animals with an adaptive advantage survived and reproduced. While advancing this theory, Darwin emphasized not only anatomical change among animals, but also the importance of the adaptation of mental processes. He looked to emotions and the expressive behaviors of emotion among animals and humans to provide support for his ideas (Plutchik, 2003). Darwin presented evidence to illustrate the similarities of purpose and expression of emotions among humans and lower animals. Examples of these expressions would include the baring of fangs among animals and the sneer of a human adult in association with fear or anger; defecation and urination in association with fear among cats, dogs, monkeys, and humans; and attempts to enlarge body size when fearful or angry (for example, bristling of hair, ruffling of feathers). Darwin felt that these types of emotional expressions evolved to fulfill a function that served the animal's ends. He proposed that these expressions might either prepare the animal for action (baring teeth prepares one to bite) or communicate intention (and hence avoid a fight).

Many theorists today accept an evolutionary explanation for the origins of emotions; this has spawned an interesting track of research that has attempted to confirm these ideas. For example, Eckman (1984) reasoned that universal recognition of emotional display through facial expression would provide evidence of common origins of emotions among humans. He conducted research to determine whether facial expressions were widely recognized across cultures. This research has shown that the facial expressions associated with *primary* emotions (e.g., happiness, sadness, anger, surprise, disgust, fear) are recognized across cultures. While these findings are not without debate, their conclusions are generally accepted to support the notion that an element of emotional response is not a cultural phenomenon, rather it is biologically based.

James and the Emphasis on Physiological Response

Another early theorist whose theoretical tradition persists to current times is William James (1884). His approach emphasized the physiological responses of emotion. James proposed that our feelings of emotion follow our physiological responses to a stimulus. He stated, for example, "We feel sorry because we cry, angry because we strike, afraid because we tremble. And not that we cry, strike, or tremble because we are sorry, angry or fearful..." (James, 1884, p. 190). Examples of empirical support were derived from research that looked at the emotional experiences of people with spinal cord injuries. This research explored the assumption that without visceral signals to the brain, emotional response would be suppressed. Another area of research that supported this view examined the extent to which facial expression (Laird, 1974) or body posture (Stepper & Strack, 1993) provided feedback to people and impacts their emotional states. Further, research indicates that certain physiological responses follow facial expressions of different emotions (e.g., increased heart rate for fear and anger expressions, decreases in heart rate for disgust) (Levenson, Ekman, & Friesin, 1990). Overall, while there may be agreement that emotions are associated with physiological responses, literature reviews on this topic have highlighted the difficulty of linking distinct physiological responses to specific emotions (Barrett, Quigley, Bliss-Moreau, & Aronson, 2004).

Cognitive Theorists Emphasize the Importance of Appraisal

Magda Arnold (1960) was an early leader of a cognitive approach to emotions, an approach that emphasized the importance of appraisal in the elicitation of emotions. The concept of cognitive appraisal deals with one's perception and interpretation of a situation. It does not necessarily mean in-depth, thoughtful cognitive processing; rather it includes direct, automatic judgments.

Arnold (1960) defined emotion as the

> Felt tendency toward anything intuitively appraised as good (beneficial), or away from anything intuitively appraised as bad (harmful). This attraction or aversion is accompanied by a pattern of physiological changes organized toward approach or withdrawal. The pattern differs for different emotions. (p. 182)

While accepting the importance of physiological changes caused by emotions, this approach places a considerable emphasis on the importance of one's personal interpretation of an emotion-arousing event.

As noted by Smith and Kirby (2000), the cognitive approach suggests it is not the stimulus alone that evokes an emotion. Instead, it is what that stimulus implies for the personal goals of the individual and that individual's beliefs, expectations, and abilities. For me, the importance of appraisal is clarified in an encounter I had with a Western rattlesnake.

I was fishing in a location where snakes are common, so I was alert to the possibility of encountering one. As I made my way upstream, I decided to cross the river. As I waded to the opposite bank, I startled a rattlesnake in the brush near the river's edge. The snake, upstream from my crossing point, slithered into the river. The cross-river speed of the swimming snake and the downriver speed of the current made the snake's path intersect with mine. I was startled, but appraised my position as safe. I vividly recall a stimulating and positive reaction to the event – noticing the grace of the reptile and its beautifully symmetrical markings against a background of clear, brilliant, sparkling water. I remained motionless and elated as the snake passed within inches of my wading boots.

Upon reuniting with my fishing companions, I shared the story and they told various stories that cast doubt on the accuracy of my appraisal of safety. I realized that, given their expectations and appraisal of the event, their emotional and behavioral response to the event would have been different from mine (and I heard plenty of advice on what I should have done!).

The cognitive approach to emotion spawned much research in the 1970s and 1980s partially because it is methodologically easier to facilitate. The cognitive approach provided a foundation for the pursuit of a social constructivist approach to emotion that is discussed in the next section.

The Mutual Construction of Culture and Emotion

Social constructivists offer a culturally relative approach to the emotion concept (Cornelius, 1996). These theorists largely dismiss the notion of emotions as inherited, cross-cultural constants and focus more on how emotions are culturally bound, learned responses. Averill (1980), one of the leading social constructivists, proposed that an emotion is a *transitory social role*. As social roles, emotions are normatively driven. That is, they are determined by a culturally derived set of rules that tell the person the proper way to appraise and respond to a given situation. These rules are part of what is learned when one is socialized into a culture.

Kitayama and Markus (1994) proposed that emotions in everyday life "depend on the dominant cultural frame in which specific social situations are constructed and, therefore, cannot be separated from culture-specific patterns of thinking, acting, and interacting" (p. 4). Similarly, Mesquita (2001a) contended "Not only does culture afford the psychological tendencies to have certain emotions, but emotions also reinforce and promote culturally important concerns" (p. 240).

This approach focuses on cultural differences of emotional display, functions, and even in types of emotions across different cultures. Nussbaum (2000) illustrated this difference by comparing her mourning behavior, caused by her mother's death, to the mourning behaviors exhibited among the Ifaluk and the Balinese. The Ifaluk wail loudly, believing that they will become ill if they fail to dwell on sad thoughts. The Balinese believe they will become ill if they *do* dwell on sad thoughts and thus are happy. In comparison, Nussbaum herself felt it was important to show sadness not only out of respect for her lost mother, but also to distract herself so as not to show helplessness. One would readily conclude from examples such as these that different cultures mold the function of emotions in different ways.

Mesquita (2001) compared individualist cultures (indigenous Dutch) and collectivist culture (Surinamese and Turkish immigrants) to empirically describe how these cultures frame emotions differently and produce different response tendencies. He suggested that the differences between the two cultures are consistent with the different value orientations of the two cultures. He found that in collectivist cultures (when compared to individualist), emotions were more grounded in one's assessment of social worth or change in social worth. In another example of culturally based research, Kitayama, Markus, Matsumoto, and Norasakkunkit (1997) examined how Japanese and Americans differ in appraising situations as self-enhancing (feelings of pride) or self-criticizing (feelings of shame). Americans believed self-esteem would increase more in success situations than it would decrease in failure situations. Japanese believed the opposite. The researchers contended that these responses reinforce the cultural values in these respective countries. Americans are motivated to enhance a positive evaluation of self, which is integral to independence. Conversely, the Japanese culture emphasizes developing and maintaining healthy social relationships and motivates people to constantly reflect on their shortcomings or improvements. In sum, emotions among the Americans tend to reinforce independence while, among the Japanese, emotions reinforce interdependence.

The Subjective Experience of Emotion

Barrett et al. (2007) proposed that prior approaches to emotion have conflated causes of emotions with the conscious state of emotion. They suggested that an account of the emotion experience requires more than a specification of causes; it must also describe content.

The experience of emotion is a system-level property of the brain. The experience of emotion has multiple levels: It can be described by its neuronal activity, but it cannot be fully equated with that activity. The overall emotional experience has both neurobiological and phenomenological features. Barrett et al. (2007) concluded that the only way we can understand the conscious state of emotion is through the first-person point of view.

The experience of emotion brings together, at a specific point in time, affect, perceptions of meaning in one's world, and one's existing knowledge about emotion. These are some properties of emotional experience that help us describe it:

- An emotion experience is a mental representation consisting of memories, imaginings, or *real-time* experiences.
- Emotion experience is caused by some object or experience.
- Emotion experience is a mental representation of pleasure or displeasure referred to as core affect.
- Emotion experiences vary by arousal content. Arousal is different from affect and can be described as a feeling of being wound up or active.
- Emotion experience involves psychological appraisals of a situation's meaning in terms of the situation's novelty, whether the situation is conductive or destructive to one's goal, whether the situation is compatible with one's norms and values, and whether one perceives oneself as being the person in a position of responsibility.
- Emotion experience has relational content, i.e., it involves a position in social relations including domination, submission, social engagement, or disengagement.

This approach to emotion offers interesting directions for the study of human–wildlife relationships. It leads us to a more complex and data-rich exploration of emotional responses to wildlife. As noted by Barrett et al. (2007), a great deal can be learned by exploring the variations of an emotion that we feel, i.e., sadness we experience over the loss of a loved one differs from the sadness we experience when hearing an animal was abused. Understanding the experience of emotion as it relates to different types of wildlife and different wildlife situations, though possibly unwieldy, would assist our understanding of human–wildlife relationships.

The Functions of Emotions

The different theoretical approaches described above would pose different explanations for the functions of emotions. There appears, for example, to be an evolutionary origin in the operation and purpose of emotions. Yet there is also a learned, cultural influence that directs topics such as appraisal, emotional display, and behavioral response.

As with much of the nature-nurture debate, answers to key questions may require the integration of these approaches. Evolutionary theorists and social constructivists emphasize different functions of emotions, for example, evolutionary theorists seek explanations in terms of human survival advantage, and constructivists emphasize cultural purpose.

Keltner and Haidt (2001a) offered a useful approach to integrating the conflicting cultural and evolutionary explanations. They proposed an approach that presumes (a) social living presents humans with problems that affect their survival; (b) emotions have evolved to solve these problems; and (c) culture loosens the link between emotions and the problems they solve, finds new ways to solve the problems, and finds new ways to use the emotions.

Keltner and Haidt (2001a) explored the social functions of emotions at four different levels: the individual level, the dyadic level, the group level, and the cultural level. This approach underscores the complexity and integration of emotions in cultural-individual systems. Keltner and Haidt (2001a) contended that at the individual level, emotions inform the individual on an event or conditions that should be acted on. Emotions can also prepare the individual for response. For example, fear over seeing a bear prepares and motivates a person to run. At the dyadic level – where we are examining relationships among people – emotional expressions (a) help individuals know other's emotions, beliefs, and intentions, (b) can evoke complementary and reciprocal emotions in others, and (c) serve as incentives or deterrents for other's social behavior. For example, showing disgust over a story about the drilling of the Artic Wildlife Refuge reveals a person's orientations and invites camaraderie. In some cases, it might deter other people from revealing their value orientations.

At the group level, emotions help define group boundaries and help individuals in the group identify other group members. Emotions also help define roles and status within groups. At the cultural level, emotions are part of a process by which people assume cultural identities; they help children learn norms and values, and they perpetuate cultural ideologies and power structures.

To conclude this review of emotion theory, the most definitive statement that can be made is that, based on a set of broad definitions, emotional responses are both biologically based and culturally directed. The complexity of emotion should not deter our exploration of it because emotion remains a powerful force directing our response to wildlife. In the next section, the role of emotion is cast in the context of the cognitive model addressed in this book.

Emotions and Cognitions

While exploring the interaction of cognitions and emotions, it must be emphasized that emotions and cognitions are theoretically separate systems, i.e., emotions have an affect on behavior that is independent of thoughtful processing. Areas of the brain aroused during episodes of emotion are different than areas activated for abstract reasoning and problem solving (Greene, Nystrom, Engell, Darley, & Cohen, 2004). More generally, Zajonc (2000) emphasized the differences between emotions and cognition on many important criteria. For example, a limited number of emotions are universal across cultures. By contrast, an infinite number of cognitions vary across cultures. Further, although we share emotions with lower animals, our cognitive processes are quite distinct from theirs. Although they are different, cognitive processes and emotional processes interact constantly in everyday life.

Effects of Emotion on Memory

The effects of emotion on memory have received attention in the literature due to the importance of memory processes to human functioning. As noted by Parrott and Spackman (2000), memory is a component of all thinking – including perception, social judgment, and problem solving that relies on recalled information. There are three ways emotions can affect memory: as a quality of what is remembered, as a condition of the mental state of an individual when encoding information, and as the condition of the individual recalling information. Research illustrates the importance of emotions in strengthening these processes. For example, research shows that memory for an emotional event is better than for an emotionally neutral event (Philippot & Schaefer, 2001). Moreover, memory for intense emotional experience shows better recall for central aspects of the event and weaker recall for background details (Heuer & Reisberg, 1992).

Other research emphasizes the importance of one's current emotional state and the effect that has on a person's recall. Early research on this topic showed mood-congruent recall; i.e., if a person is in a positive state, that person tends to recall memories that are also positive (Bower, 1981). However, since that time, research has emphasized the importance of the person's motivations during recall, the person's personality characteristics, and the situation in which recall occurs. This suggests, for example, that people will evoke memories that are consistent with their goals: A person preparing for an elk hunt will recall memories of prior positive hunts. The situation in which recall occurs also has an important impact on what is recalled and helps explain the occurrence of mood-incongruent recall. Finally, research has shown that some individuals are better at emotional *self-regulation* and that these individuals may be better at controlling the recall of memories; i.e., they may be able to recall happy memories when they are sad, hence balancing their emotional state (Parrott & Spackman, 2000).

Emotion's Effects on Decision Processes

While it was once believed that rational decisions should be void of emotions, current views emphasize the critical importance of emotions in rational-decision processes. As recently as the mid-1990s, Damasio (2005) advanced a controversial view that emotions played an important role in effective decision making and that reasoning systems actually evolved as an extension of the autonomic emotional system. Damasio (2005) developed the somatic marker hypothesis that suggests emotions mark certain aspects of a situation or certain outcomes of possible actions which would be experienced as a *gut feeling*. Damasio based his theory on his work as a neurologist. He observed that patients who suffered brain damage and whose ability to experience emotions were impaired, made poor, often self-destructive, decisions.

Recent reviews of research on this topic show that the affective or emotional state of individuals has a significant effect on their judgments (Forgas, 2003; Isen, 2000; Petty, Fabrigar, & Wegener, 2003). Emotional state affects people's judgments about themselves, about others, and about the nature of their social interactions. Findings generally show *affect consistency* in evaluations; e.g., a person in a positive emotional state is more likely to evaluate something as positive. Findings also indicate that when people are induced with positive affect, they are more creative or innovative in their responses and more likely to believe that they will succeed (Isen, 2000); those induced with negative affect engage in effortful, analytic, and vigilant processes (Forgas, 2003). Moreover, with positive affect present, people make risk-averse decisions when the probability of loss is real and meaningful (Arkes, Herren, & Isen, 1988), but are more risk-prone when the probability of loss is less significant (Isen & Patrick, 1983). Decision accuracy also appears to be impacted by positive affect. For example, a study of medical students showed that students induced with positive affect made correct diagnosis sooner than those not induced and that positively affected students also went beyond the assigned tasks when considering possible treatments (Isen, Rosenweig, & Young, 1991).

Affect appears to facilitate constructive group decision processes (Forgas, 2003). Carnevale and Isen (1986) conducted a study that showed positive affect can facilitate negotiation in an *integrative-bargaining situation*. This bargaining task poses a situation in which the optimal agreement requires tradeoffs on issues of importance. Findings showed that people in the positive-affect condition were less likely to break an agreement and more likely to reach the optimal agreement. They were also less likely to use aggressive negotiating tactics and reported more enjoyment in the task. Similarly, Forgas (1998) found that the induction of positive mood led to the formulation of more optimistic, cooperative, and integrative negotiating strategies as well as more positive attitudes toward negotiating partners.

These findings raise important questions for researchers and practitioners alike. What forms of conflict resolution and stakeholder-engagement might be devised that explore the utility of creating positive affect as a base for more effective and lasting compromise?

While emotions, and affect more generally, have an important effect on decisions, it appears that the role of emotions in decision making varies considerably depending on the nature of the decision task. Forgas (2000) proposed that one's affective state is particularly important in influencing decisions involving deep deliberation or that are based on heuristic decision rules. He contended, however, that affect has little impact on decision processes where judgments are already crystallized.

Perhaps more relevant for examining human–wildlife relations is Greene et al.'s (2004) proposal that emotion-based processing is more prevalent in personal moral decisions, while in-depth cognitive processing is more prevalent in impersonal moral decisions. *Personal moral decisions* are those that involve direct harm to another. Greene et al. specified this as harm to other people, but one might readily expand that to include harm to other living beings. A personal moral decision involves agency of the person deliberating (I, not someone else, is inflicting the harm), and the decision is not the result of the agent trying to protect another living being. To illustrate, a personal moral decision might include a situation where a person finds a wounded animal and faces the decision of ending the animal's life to end its suffering.

If the decision does not meet these criteria, it would be considered an *impersonal moral decision*. An impersonal moral decision might involve a person issuing an order to shoot many animals of a specific species (e.g. non-indigenous goats on the Galapagos) to protect the habitat of other species. In this case, other people inflict harm and the harm is inflicted to protect another species.

Green et al. presented evidence that these two types of decisions appear to stimulate different areas of the brain; personal moral decisions stimulate areas related to social cognition and emotion, while impersonal moral decisions stimulate areas related to in-depth cognitive processing. Greene et al. (2004) theorized that response to personal moral dilemmas has deep evolutionary roots. Our early ancestors lived social lives where they would be presented with such decisions, and their action would be dictated by emotion, not by moral reasoning.

In modern humans, such personal moral judgements would be, as proposed by Haidt (2001), rapid, intuitive, emotion-based, and justified later (due to social demands) through cognative processes. Impersonal moral judgments are usually driven by in-depth cognitive activity. Most interestingly, these two processes would interact and even compete in some cases of moral reasoning, such as when a person is confronted with overriding or controlling a personal moral.

The distinction of personal versus impersonal moral judgment merits further exploration in examining human–wildlife relationships. It may be useful to explore positions on wildlife treatment based on how people frame the moral situation. Are groups that differ on wildlife values or on wildlife issues more similar in their personal moral judgments than impersonal judgments? Are impersonal judgments more susceptible to cultural differences and culture shift? Further, can we explore apparent inconsistencies in people's attitudes or wildlife value orientations based on this distinction?

For example, research by Manfredo and Teel showed that a large percentage of residents in the western United States hold conflicting value orientations (high on both mutualist and domination orientations). Perhaps this is a reflection of conflict among their moral judgments (e.g., I would never go hunting because I could not shoot an animal, but I support my spouse going hunting). Many situations that deal with wildlife provide a contrast to these types of judgments.

Emotions and Attitude Change

Research and practical applications suggest that emotional arousal can increase the effectiveness of persuasion attempts (Cafferata & Tybout, 1989; Dillard & Meijnders, 2002; Petty & Caccioppo, 1981; Nabi, 2002; O'Shaughnessy &

O'Shaughnessy, 2003). In advertising, for example, emotional appeals are a unique style that is believed to be effective in specific circumstances (Calder & Gruder, 1989).

Research suggests that emotion-inducing appeals can be effective; however, contextual factors make simple generalizations elusive. For example, Zinn and Manfredo (2000) tested whether sadness emotional appeals were more effective than rational appeals in swaying people's position on the vote to ban trapping in Colorado. Although prior research suggests that messages evoking anger or sadness lead to attitude change, Zinn and Manfredo found that the emotional appeals (including pictures of animals caught in traps), although more memorable, were no more effective than rational appeals. This is not a definitive finding, but it illustrates a common inconsistency found in persuasion research.

In a review of the role of discrete emotions on persuasion, Nabi (2002) proposed the effects of emotion occur in one of three ways. First, emotions can serve as heuristics or rules of thumb that guide thought with minimum cognitive processing. Nabi concluded this occurs under conditions of extreme emotion arousal or when positive emotions are involved. For example, a violator stopped by a game warden may experience relief upon receiving only a warning and, at that point, would be susceptible to attitude change (simply because the violator is told to change).

Second, emotions can stimulate selective cognitive processing. This is consistent with the more commonplace view of how persuasion works – give information to people, and they will think about the information and change if the arguments are accepted. In this case, the experience of the emotion creates a goal (e.g., fear might evoke escape or avoidance goals), and the greater the intensity of the emotion, the more likely information processing will occur. For example, if a person about to hike on a certain trail is told a bear mauled a hiker in that area the day before, that person, now fearful, would be attentive to information that would alleviate the fear; this person would probably engage in highly deliberative processing: take new information, link it to pre-existing information, and arrive at a new attitudinal position (other places to go, bear avoidance information).

Third, emotions promote selective processing by creating a *frame* through which certain types of information become more salient than others. Research suggests that when people are in a happy state, they tend to be more likely to attend to and remember positive aspects of an appeal and be more likely to change in a manner consistent with the positive state (Dillard & Meijnders, 2002).

Certainly this area of investigation merits much further attention by researchers. The topic of wildlife is closely aligned to the development of emotions within humans. That is, humans readily relate to wildlife at an emotional level. The use of emotion-evoking strategies holds much promise in developing effective means of communication.

Emotions and Norms

Norms are linked to emotions in two ways. First, there are actual norms about displaying emotional reactions. Second, emotion is aroused when social norms or moral standards are violated.

According to Fischer et al. (2004), the most common approach for examining why people regulate their emotions is based on their perception of the discrepancy between the emotion experienced and the emotional norm for a situation. So, for example, a manager at a public meeting who is being verbally abused may feel a welling of anger; however, the manager strives to hide this anger because of a workplace norm (i.e., do not react emotionally to stakeholders). Interestingly, research suggests that those who must exhibit a high degree of emotional regulation in their occupation may also have high job dissatisfaction if they feel the role they play is inconsistent with their role identity. This might happen when wildlife managers must make decisions that support the political regime when they realize the decision violates norms of the wildlife profession. More broadly, Fischer et al. suggested that people regulate their emotions for three basic reasons: (a) impression management (i.e., to avoid being evaluated negatively by others), (b) pro-social motives (i.e., enhancing, protecting, or not hurting others), and (c) to influence the behavior of others.

The topic of emotion norms has interesting implications for organizational development in wildlife agencies and non-governmental agencies. For example, what emotion norms are held by professionals who deal frequently with stakeholders? To what extent can norms be reshaped and to what extent should the ability to regulate emotions (in dealing with publics) be used in hiring practices?

While people are guided by emotion norms, emotions also guide behavior as a sanction applied when norms are violated. For example, the display of anger or disgust in response to a normative violation and shame or embarrassment when receiving the admonishment helps preserve an accepted norm. Moreover, display of emotion serves to identify those who do not hold the norm and who are not members of a particular group. For example, a wildlife photographer might tell an associate photographer that she set out bait to attract a species she wanted to photograph. The friend might display disgust over the situation, causing the photographer to feel shame. This would be tied to norms regarding the depiction of natural conditions in one's photographs (or even being truthful in what an image shows), which are linked to one's motives or goals of being a good conservationist. Another person who does not identify with this social group may see little distinction between these behaviors (i.e., he may not endorse the norms or respond to the emotional display).

Moral emotions include shame, guilt, and pride; they arise (or are anticipated) when people judge that their actions are against their own moral standards (Tangney, Stuewig, & Mashey, 2007). When people transgress, they usually feel shame, guilt, or embarrassment, and when they do good deeds they feel pride and self-approval. Moral emotions provide an important motivational force for individuals to act in concert with the interests and welfare of others (Haidt, 2003).

An exploration of what types of actions arouse moral emotions will help us understand people's actions toward wildlife and the moral boundaries of human–wildlife relationships for individuals and societies. Emotions such as righteous anger, disgust, or contempt toward the actions of others help explain an individual's behavior toward other people who have offended a moral standard. Conversely, an exploration of incidents creating shame and guilt helps us understand the individual's own self-evaluation processes and possible adaptive behavior. Further, an exploration of coping mechanisms will aid in understanding how people minimize these emotions and justify actions believed to be morally wrong. For example, research by Frommer and Arluke (1999) showed that those who surrender their pets to shelters feel guilt (regret and doubt) as do those who work in animal shelters and are involved in euthanizing animals. Coping strategies are used for dealing with this guilt. Those who surrender animals tend to blame someone else (landlord, spouse, parent), pass the buck (emphasizing the adoptability of the animal or making the shelter responsible), or blame the victim (keeping other animals and humans safe from the animal). Workers also tend to blame others (those who surrender animals) or the victim (a better alternative to bad life or painful death). A fruitful line of inquiry would explore differences among people who consider a specific behavior toward wildlife as immoral but justifiable versus those who simply believe the behavior is not immoral.

A second area of theory that would be useful, as suggested by Hill (1993), would be exploring how moral empathy motivates human–wildlife relationships. Tangney, Stuewig, and Mashek (2007) described empathy as a moral emotional process in which the individual shares an emotional response with another person. While this definition does not explicitly include sharing emotional responses with other animals, I have noted elsewhere in this book that, due to anthropomorphic tendencies, humans project human characteristics onto animals, including emotions. The extent to which one empathizes toward animals would dictate the extent to which that person would want to exhibit helping behavior to alleviate emotional distress. This approach is supported by Hill (1993) who tested the role of empathy as the motivational basis of attitudes toward animals. In a sample of farmers and animal rights activists in Australia, she found animal rights activists had a strong identification base and a low instrumental base for attitudes toward animals, while farmers had the opposite scoring. Poresky (1990) developed a scale for measuring children's empathy toward animals and found that children's empathy toward animals was related to their empathy toward other children.

Emotions and Values

A critical element of the cognitive approach to emotions is the notion of goals. Emotions are the result of an individual's appraisal with respect to the goal that individual pursues. In this context, goals and their origins shape emotional response. It is a logical conclusion that values, which are superordinate goals that guide social life, are linked to the arousal of emotion. This view reflects the essence of Keltner and Haidt's (2001b) suggestion that throughout human history, survival-based functions of emotions have been replaced by culturally based reasons (e.g., maintaining group solidarity, minimizing group defection). It explains why different people observing the same event may have vastly different emotional responses. Emotional differences are rooted in differences in value orientations.

Fischer, Manstead, and Mosquera (1999) provided support for this explanation in a study that explored the relationship between values and emotional prototypes in a cross-cultural context. Using Schwartz's measurement instrument, they first established value priority differences between the Netherlands and Spain. Results showed that the Dutch placed greater emphasis on individualistic values such as ambition, capability, freedom, independence, and self-discipline. The Spanish group placed greater emphasis on collectivistic values including respect for tradition, respect for parents and the elderly, social power, and family security. In a second stage of the study, subjects were asked to describe typical emotional episodes for three key emotions: pride, shame, and anger. Tests of hypotheses reinforced the value–emotion consistency within cultures and differences between cultures. For example, the antecedents of pride and shame were more likely to be focused on self-related appraisal and personal achievement for the Dutch and community-related appraisal for the Spaniards.

Emotions and Wildlife Value Orientations

Emotional responses to wildlife are closely related to value orientations. Consider the most likely situations involving wildlife in modern society: hearing or reading news stories about a wildlife issue or incident, seeing wildlife while driving, watching a television program about wildlife, seeing wildlife while on a recreational trip, seeing evidence around home where wildlife have caused damage, etc. Our emotional responses to these events are dictated largely by situational specifics and our goals at the time. Individuals with different goals can have vastly different emotional responses. Consider, for example, the response of two different individuals who hear the story of a rancher who traps a coyote, kills it, and hangs the carcass on a fence. Without additional information, one individual may express sadness and disgust while another may express a neutral or satisfied emotion. These emotional responses might reflect different motivations (eliminating cruel treatment of animals versus economic production) or different wildlife value orientations (e.g., the former individual having a mutualism wildlife value orientation while the latter having a domination wildlife value orientation). These emotional responses might change with subtle alterations in the situation. For example, if the rancher does not hang the carcass, emotional intensity may diminish. A similar effect would be found if it was determined that the coyote had attacked a group of children (evoking other goals). In summary, many emotional responses to wildlife are tied to values and can be fruitfully explored in that context.

An initial area of exploration is with cognitive-emotional consistency evident in responses to wildlife. We would expect that there are distinct situations that evoke predictable emotional responses dependent upon a person's value orientation. For example, we might expect anger and disgust from people with mutualism orientations who hear stories about hunters pursuing trophies, affiliation emotions for newborn animals among those with caring value orientations, and anger among those with domination values when there are discussions of passing laws against hunting.

Emotions enforce and reinforce the values and norms important to a particular social group. In addition, the emotion can communicate a basis for social acceptance or rejection. For example, a display of disgust over a given wildlife issue, such as wildlife trapping, conveys a person's orientation. This display invites response from others. It provides the basis of acceptance or rejection, commonality or difference, approach or withdrawal from the individual. The display helps define social group boundaries.

The link between emotions and value orientations has interesting implications for research on human–wildlife relationships. First, research should explore whether there are predictable relationships among specific situations, value orientations, and emotional responses. A key consideration in this exploration would determine the ways in which situations are classified. Most immediately, it would make sense to classify situations by the extent to which they would threaten or offend a given value orientation. Beyond the exploration of that relationship, an important question would ask the extent to which emotional intensity evokes certain forms of behavior. An important question might be: To what extent does emotion and its intensity explain behavioral response over and above traditional attitude measures?

Finally, if the relationship between wildlife value orientations and specific emotional responses is predictable, then measures of emotions would provide useful indicators of people's wildlife value orientations. This might lead to development of an assessment tool that measures emotional response to a specific scenario and which allows inference to specific value orientations. This approach might yield an easy-to-administer and parsimonious method of assessing wildlife value orientations.

Emotions, Health Effects, and Interactions with Wildlife

Research shows that negative emotions such as anger, fear, and distress are associated with morbidity and mortality in almost every form of chronic illness (Mayne, 2001). While researchers debate whether these emotions are a cause, effect, or merely an indicator of negative consequences (Leventhal & Patrick-Miller, 2000), they believe that these emotions have a contributory factor. While it is unclear whether positive emotional states can lead to long-term health benefits, researchers speculate that benefits are likely.

Although the effects of positive emotions have received less attention in the literature, they pose interesting implications for human–wildlife relationships. Most notable may be the possible therapeutic emotional effects that result from human interaction with animals, primarily pets. Serpell (2003) suggested, in reviewing the benefits of pet ownership, that human health improves from the social support offered by the human–animal relationship. Vining (2003) summarized these effects and suggested that animals can

- Reinforce human self-worth through what is perceived as the animal's unconditional love for the human.
- Help the human develop a sense of self and self-esteem, offer comfort, companionship, and social support.

These positive effects may improve health. Friedman, Katcher, Lynch, and Messent's (1980) study showed that pet companionship correlated with successful recovery from heart attacks.

Another area of speculation involves the effects of positive emotions associated with play and leisure. Recreation involving wildlife (e.g., hunting, fishing, viewing) is a popular pursuit in post-industrialized nations. Increasingly, researchers are recognizing the benefits of wildlife-related recreation and other forms of leisure; this has led researchers to promote a *benefits-based management approach* for practitioners (Driver & Bruns, 1999). The positive emotional states associated with leisure experiences are a component of the benefit.

The emotional benefit of recreation is supported by Fredrickson and Branigan (2001), who proposed that play creates a unique opportunity where benefits can arise through positive emotions. They suggested that joy (or happiness) inspires play and that play has these benefits: building friendships and attachments, promoting skill acquisition, developing cognitive skills, and fostering creativity and innovation. They suggested that interest is a basic emotion and that this emotion is associated with leisure experiences. They contended that interest promotes exploration, extension of the self, and acquisition of new information. Interest is associated with feeling animated and enlivened, which encourages an openness to new ideas, experiences, and actions.

Fredrickson and Branigan (2001) contended that the concept of *flow*, or optimal experience, introduced by Csikszentmihalyi (1990), is an extension of interest. Csikszentmihalyi suggested that flow occurs when the degree of challenge perfectly matches a person's ability, a situation that occurs frequently in leisure. Flow can occur for the fly angler who becomes lost in his activity, loses all self-consciousness, is unaware of the passage of time, and concentrates only on the event. Csikszentmihalyi (1990) suggested that flow experience is a characteristic of achieving self-actualization and higher levels of physical and mental health.

The positive effects on human experience via emotions is an under-explored topic. This is partly due to psychology's focus on problems versus benefits (Seligman & Csikszentmihalyi, 2000; Fredrickson & Branigan, 2001). Because of their role in aggression, violence, depression, and suicide, negative emotions have received more attention; however, the body of research exploring the positive consequences of human experience is growing. Given the attention-focusing, inherited response that people have toward wildlife, positive emotions will be a useful area for further investigation (Katcher & Wilkins, 1993).

Summary

Elster stated "Emotions matter because if we did not have them nothing else would matter" (1999, p. 403). A rancher angrily berates a wildlife manager because a decision affects the rancher's traditional livelihood; a woman dotes lovingly over a fawn that was found abandoned; a fly angler becomes one with his surroundings as the rising fish match his fly presentation abilities and he is lost in flow experience; a family grieves the loss of a son who has been attacked by a mountain lion. It is the emotion of these events that make them matter to us. That fact alone should compel us to increase research on emotional responses to wildlife.

The review of emotion presented here suggests that emotions are an incredibly complex topic. Researchers have multiple views on the concept, ranging from its physiological associations to the relativity of its cultural meaning. Primary emotions are an inherited response linking us to others in the animal world. However, primary emotions are also culturally molded, serving functions that preserve norms and values of social groups.

Emotions act with cognitions to direct human behavior. They play an important role in memory, decision making, and attitude change; they clarify roles and social structure. Emotions are also linked to health benefits, an interesting implication for those exploring human–wildlife relationships.

Management Implications

Wildlife professionals should re-examine the widely held view that emotional response issues are trivial, unimportant, or non-informative. Emotional responses are a barometer of ideals that are deeply important to people and an important form of communication when management agencies deal with publics (Vining & Tyler, 1999). Emotional displays frequently signify that something important is at stake to participants. More specifically, emotions reveal implications regarding threats to (or reinforcement of) people's identities, their values, and their norms. Emotions merit careful consideration and thoughtful response.

Emotions and decision-making. The profession is constantly seeking and developing prescriptive approaches to improving decision making in natural resources. The prevailing paradigm in these approaches depicts a highly cognitive, rational approach. Yet findings in the emotions literature suggest that emotion and affect are critical to effective decision making for individuals. This does not mean a rational approach should be abandoned; however, we should not ignore our intuition when making decisions. We should recognize the legitimacy of a "gut feeling" about a decision (as it is proposed to be an affective signal) and encourage establishing structured ways to accommodate input at that level.

Emotions and collaboration. The prevailing emotion in a situation affects interpersonal behavior. The prevailing emotional state will affect our ability to communicate with others, to achieve stakeholder consensus, and to reach conflict resolution. Negative affect inhibits these outcomes. As we structure our interactions with stakeholders, an important first step is to establish a positive affective state prior to negotiations. This might be accomplished by focusing on areas of agreement, facilitating social engagement to make a person feel accepted, eliminating physical barriers that would separate a manager from stakeholders, etc.

Emotions and persuasion. Evoking emotional responses to our communication efforts can increase the effectiveness of persuasive appeals. Natural resource professionals often communicate with the public in a highly factual, cognitive fashion, yet people relate strongly to wildlife at an emotional level. We could improve our communication by developing strategies that evoke emotional reactions. Given the notorious difficulties with persuasion attempts, this requires careful formative analysis and pilot testing. But this approach might improve communication efforts considerably.

Emotions and attracting and retaining professionals. Increasingly, we are challenged to attract and retain natural resource professionals. The job of a natural resource manager can be highly demanding with sustained levels of daily conflict. Our training of fish and wildlife professionals and our attempts to improve organizations should deal directly with the topics of emotional norms and emotional intelligence. Employees should be aware of emotional norms and how to cope with the challenges to these norms. Also, employees should improve their ability to recognize other's emotional communication (emotional intelligence) because this facilitates better communication, management, and stakeholder relations.

Current discussions about managing natural resources focus on recognizing the services they provide. Positive emotional response associated with wildlife contact benefits human health. We should not underemphasize these benefits to humans simply because they do not have a clear utilitarian purpose.

Summary Points About Emotion

- Emotion is part of affect, or the feeling states, of individuals. Mood is the ongoing background affective state, while emotion is an event-specific spike of affect. Just a small number of emotions appear to be universal to humans.

- There has been preoccupation with cognitive approaches to human behavior; however, this is changing rapidly. The long-held notion that emotion leads to unproductive decisions is shown inaccurate. Emotions and affect are critical to effective human functioning.
- Emotion is examined from many perspectives: its expressive reactions, physiological responses, its importance to appraisal, which gives rise to emotion, its cultural influences, and its actual subjective experience. All of these perspectives illuminate a complex human process.
- Emotional responses are tied to our evolutionary roots. They are responses we share with other living beings. Emotions developed through selection processes to enhance human survival (e.g., preparing a human to run); as humans became societal and cultural beings, emotions sustained cultural forms.
- Emotions and cognitions are separate systems and are linked to different brain areas; when people deliberate, emotions and cognitions interact.
- Emotions impact what we remember and store in memory. Emotions can be retrieved as information, and they help us anticipate and predict future events.
- People's emotional states and the judgments interact (e.g., positive emotion leads to more positive evaluations). Positive emotions lead to more effective decisions and facilitate more cooperative inter-group functioning.
- Some decisions evoke more emotionally based processes than others. For example, impersonal moral decisions may be linked to cognitive processing, while personal moral decisions may have an emotional base.
- Persuasive appeals that evoke emotion can be highly effective, though somewhat contextually dependent.
- Norms govern our display of emotions; norms dictate what is appropriate to reveal about our feelings. Moreover, emotions operate as sanctions; they ensure normative behavior is followed (e.g., a dirty look from others is a sanction applied to the person who violates a norm). Moral emotions (shame, guilt, pride) and their associated moral standards may be particularly relevant as we examine people's treatment of wildlife.
- Emotional responses are strongly linked to values and value orientations. People's emotional responses to an event will differ depending on how the event is appraised relative to their own goals (i.e., values).

References

Arkes, H. R., Herren, L. T., & Isen, A. M. (1988). Role of possible loss in the influence of positive affect on risk preference. *Organizational Behavior and Human Decision Processes, 42*, 181–193.

Arnold, M. (1960). *Emotion and personality: Psychological aspects* (Vol. 1). New York: Columbia University Press.

Averill, J. R. (1980). A constructivist view of emotion. In R. Plutchik, & H. Kellerman, (Eds.), *Emotion: Theory, research and experience* (Vol. 1, pp. 305–339). New York: Academic Press.

Barrett, L. F., Quigley, K., Bliss-Moreau, E., Aronson, K. R. (2004). Arousal focus and interoceptive sensitivity. *Journal of Personality and Social Psychology, 87*, 684–697.

Barrett, L. F., Mesquita, B., Ochsner, K. N., & Gross, J. J. (2007). The experience of emotion. *Annual Review of Psychology, 58*, 373–403.

Bower, G. H. (1981). Mood and memory. *American Psychologist, 36*, 129–148.

Cacioppo, J. T., & Gardner, W. L. (1999). Emotion. *Annual Review of Psychology, 50*, 191–214.

Cafferata, P., & Tybout, A. (1989). *Cognitive and affective responses to advertising*. Lexington, Mass: Lexington Books.

Calder, B. J., & Gruder, C. L. (1989). Emotional advertising appeals. In P. Cafferata, & A. Tybout (Eds.), *Cognitive and affective responses to advertising* (pp. 277–285). Lexington, Mass.: Lexington Books.

Carnevale, P. J. D., & Isen, A. M. (1986). The influence of positive affect and visual access on the discovery of integrative solutions in bilateral negotiation. *Organizational Behavior and Human Decision Processes, 37*, 1–13.

Cornelius, R. (1996). *The science of emotion: Research and tradition in the psychology of emotion*. Upper Saddle River, NJ: Simon and Schuster.

Csikszentmihalyi, M. (1990). *Flow: The psychology of optimal experience*. New York: Harper Perennial.

Damasio, A. (2005). *Descartes' error: Emotion, reason and the human brain*. London: Penguin Books.

Dillard, J. P. & Meijnders, A. (2002). Persuasion and the structure of affect. In J. P. Dillard, & M. Pfau (Eds.), *The persuasion handbook: developments in theory and practice* (pp. 309–328). Thousand Oaks, CA: Sage Publications.

Driver, B. L., & Bruns, D. (1999). Concepts and uses of the benefits approach to leisure. In E. L. Jackson, & T. L. Burton (Eds.), *Leisure studies: Prospects for the twenty-first century*. State College, PA: Venture Publishing, Inc.

Elster, J. (1999). *The alchemies of the mind*. Cambridge: Cambridge University Press.

Eckman, P. (1984). Expression and the nature of emotion. In K. Scherer, & P. Eckman (Eds.), *Approaches to emotion* (pp. 319–343). Hillsdale, NJ: Erlbaum.

Fischer, A. H., Manstead, A. S. R., & Mosquera, P. M. R. (1999). The role of honor-based versus individualistic values in conceptualizing pride, shame, and anger: Spanish and Dutch cultural prototypes. *Cognition and Emotion, 13*, 149–179.

Fischer, A. H., Manstead, A. S. R., Evers, C., Timmers, M., & Valk, G. (2004). Motives and norms underlying emotion regulation. In P. Philippot, & R. S. Feldman (Eds.), *The regulation of emotion* (pp. 187–210). Mahwah, NJ: Laurence Erlbaum Associates.

Forgas, J. P. (1998). On feeling good and getting your way: Mood effects on negotiation strategies and outcomes. *Journal of Personality and Social Psychology, 74*, 565–577.

Forgas, J. P. (2000). Affect and information processing strategies: An interactive relationship. In J. P. Forgas (Ed.), *Feeling and thinking: the role of affect in social cognition* (pp. 253–282). Paris: Cambridge University Press.

Forgas, J. P. (2003). Affective influences on attitudes and judgments. In R. J. Davidson, K. R. Scherer, & H. H. Hill (Eds.), *Handbook of Affective Sciences* (pp. 596–618). Oxford: Oxford University Press.

Fredrickson, B. L., & Branigan, C. (2001). Positive emotions. In T. J. Mayne, & G. A. Bonanno (Eds.), *Emotions: current issues and future directions* (pp. 123–151). New York: Guilford Press.

Friedman, E., Katcher, A., Lynch, J., & Thomas, S. (1980). Animal companions and one year survival of patients discharge from a Coronary Care Unit. *Public Health Reports, 95*, 307–312.

Frommer, S. S., & Arluke, A. (1999). Loving them to death: blame-displacing strategies of animal shelter workers and surrenderers. *Society and Animals, 7*(1), 1–16.

Greene, J. D., Nystrom, L. E., Engell, A. D., Darley, J. M., & Cohen, J. D. (2004). The neural bases of cognitive conflict and control in moral judgment. *Neuron, 44,* 389–400.

Haidt, J. (2001). The emotional dog and its rational tail: a social intuitionist approach to moral judgment. *Psychological Review, 108,* 814–834.

Haidt, J. (2003). Elevation and the positive psychology of morality. In C. L. Keyes, & J. Haidt (Eds.), *Flourishing: Positive psychology and the life well-lived* (pp. 275–289). Washington, DC: American Psychological Association.

Heuer, F., & Reisberg, D. (1992). Emotion, arousal, and memory for detail. In S. A. Christianson (Ed.), *The handbook of emotion and memory: Research and theory* (pp. 151–180). Hillsdale, NJ: Erlbaum.

Hill, A. M. (1993). The motivational bases of attitudes toward animals. *Society and Animals, 1*(2), 111–128.

Isen, A. M. (2000). Positive affect and decision making. In M. Lewis, & J. M Haviland-Jones (Eds.), *Handbook of emotions* (2 nd ed., pp. 417–435). New York: The Guilford Press.

Isen, A. M., Rosenweig, A. S., & Young, M. J. (1991). The influence of positive affect on clinical problem solving. *Medical Decision-Making, 11*(3), 221–227.

Isen, A. M., & Patrick, R. (1983). The effect of positive feelings on risk-taking: When the chips are down. *Organizational Behavior and Human Performance, 3,* 194–202.

Izard, C. E. (1977). *Human emotions.* New York: Plenum Press.

James, W. (1884). What is an emotion? *Mind, 19,* 188–205.

Katcher, A., & Wilkins, G. (1993). Dialogue with nature: Its nature and culture. In S. R. Kellert, & E. O. Wilson (Eds.), *The biophilia hypothesis* (pp. 173–199). Washington, DC: Island Press.

Keltner, D., & Haidt, J. (2001a). Social functions of emotions. In T. J. Mayne, & G. B. Bonanno (Eds.), *Emotions: Current issues and future directions* (pp. 192–212). New York: Guilford Press.

Keltner, D., & Haidt, J. (2001b). Social functions of emotions at four levels of analysis. In W. G. Parrott (Ed.), *Emotions in social psychology* (pp. 175–184). Philadelphia, PA: Taylor Francis.

Kemper, T. (1987). How many emotions are there? Wedding the social and autonomic components. *American Journal of Sociology,* 93, 263–289.

Kitayama, S., & Markus, M. R. (Eds.). (1994). *Emotion and culture: empirical studies of mutual influence.* Washington, DC: American Psychological Association.

Kitayama, S., Markus, H. R., Matsumoto, H., & Norasakkunkuit, V. (1997). Individual and collective processes in the construction of the self: Self-enhancement in the United States and self-criticism in Japan. *Journal of Personality and Social Psychology, 72,* 1245–1267.

Laird, J. D. (1974). Self-attribution of emotion: the effects of expressive behavior on the quality of emotional experience. *Journal of Personality and Social Psychology, 29,* 475–486.

Levenson, R. W., Ekman, P., & Frisen, W. V. (1990). Voluntary facial expression generates emotion-specific nervous system activity. *Psychophysiology, 27,* 363–384.

Leventhal, H., & Patrick-Miller, L. (2000). Emotions and physical Illness: causes and indicators of vulnerability. In M. Lewis, & J. Haviland-Jones (Eds.), *Handbook of Emotions* (2 nd ed., pp. 523–537). New York: Guilford Press.

Mayne, T. J. (2001). Emotion and health. In T. J. Mayne, & G. A. Bonanno (Eds.), *Emotions: current issues and future directions* (pp. 361–397). New York: Guilford Press.

Mesquita, B. (2001a). Culture and emotion. In T. J. Mayne, & G. A. Bonanno (Eds.), *Emotions: current issues and future directions* (pp. 214–250). New York: Guilford Press.

Mesquita, B. (2001b). Emotions in collectivist and individualist contexts. *Journal of Personality and Social Psychology, 80*(1), 68–74.

Milton, K. (2002). *Loving nature: Towards an ecology of emotion.* New York: Routledge.

Nabi, R. L. (2002). Discrete emotions and persuasion. In J. P. Dillard, & M. Pfau (Eds.), *The persuasion handbook: Developments in theory and practice* (pp. 289–308). Thousand Oaks, CA: Sage Publications.

Nussbaum, M. C. (2000). Emotions and social norms. In L. Nucci, G. B. Saxe, & E. Turiel (Eds.), *Culture, thought and development* (pp. 41–63). Mahwah, NJ: Lawrence Erlbaum Associates.

O'Shaughnessy, J., & O'Shaughnessy, N. J. (2003). The marketing power of emotion. Oxford: Oxford University Press.

Petty, R. E., & Cacioppo, J. T. (1981). Attitudes and persuasion: Classic and contemporary approaches. Dubuque, IA: Wm. Brown Co. Publishers.

Parrott, W. G., & Spackman, M. P. (2000). Emotion and memory. In M. Lewis, & J. M. Haviland-Jones (Eds.), *Handbook of Emotions* (2 nd ed., pp. 476–490). New York: Guilford Press.

Petty, R. E., Fabrigar, L. E., & Wegner, D. T. (2003). Emotional factors in attitudes and persuasion. In R. J. Davidson, K. R. Scherer, & H. H. Hill (Eds.), *Handbook of Affective Sciences* (pp. 752–772). Oxford: Oxford University Press.

Philippot, P., & Schaefer, A. (2001). Emotion and memory. In T. J. Mayne, & G. A. Bonanno (Eds.), *Emotions: Current issues and future directions* (pp. 82–122). New York: Guilford Press.

Plutchik, R. (2003). *Emotions and life: Perspectives from psychology, biology and evolution.* Washington, DC: American Psychological Association.

Poresky, R. H. (1990). The young children's empathy measure: Reliability, validity, and effects of companion animal bonding. *Psychological Reports, 66,* 931–936.

Rosenberg, E. L. (1998). Level of analysis and the organization of affect. *Review of General Psychology, 2,* 247–270.

Seligman, M. E. P., & Csikszentmihalyi, M. (2000). Positive psychology: An introduction. *American Psychologist, 55,* 5–14.

Serpell, J. A. (2003). Anthropomorphism and anthropomorphic selection – Beyond the "Cute Response." *Society and Animals, 11,* 183–100.

Smith, C. A., & Kirby, L. D. (2000). Consequences requires antecedents: Toward a process model of emotion elicitation. In J. D. Forgas (Ed.), *Feeling and thinking: The role of affect in social cognition* (pp. 83–106). New York: Cambridge University Press.

Stepper, S., & Strack, F. (1993). Proprioceptive determinants of emotion and nonemotional feelings. *Journal of Personality and Social Psychology, 64,* 211–220.

Tangney, J. P., Stuewig, J., & Mashek, D. J. (2007). Moral emotions and moral behavior. *Annual Review of Psychology, 58,* 345–372.

Vining, J., & Tyler, E. (1999). Values, emotions and desired outcomes reflected in public responses to Forest management plans. *Human Ecology Review, 6*(1), 21–34.

Vining, J. (2003). The connection to other animals and caring for nature. *Research in Human Ecology, 10*(2), 87–99.

Zajonc, R. B. (2000). Feeling and thinking. In J. P. Fargas (Ed.), *Feeling and thinking: the role of affect in social cognition* (pp. 31–58). Paris: Cambridge University Press.

Zinn, H. C., & Manfredo, M. J. (2000). An experimental test of rational and emotional appeals about a recreation issue. *Leisure Sciences, 22,* 183–194.

Chapter 4
Attitudes and the Study of Human Dimensions of Wildlife[1]

Contents

---

[1] This chapter was co-authored with Alan D. Bright.

Introduction

What role do alternative regulations have in stimulating or depressing water-fowl hunting participation? A group of managers entrusted with the regulation of recreational waterfowl hunting in North America examined this question (Case, 2004). The potential waterfowl regulations would dictate things such as season length and species bag limit. Group discussions focused on the attitu-dinal studies that could guide selection of regulations. The managers proposed an attitude survey. In this survey, hunters would be asked a series of questions to assess thier support for or opposition to different regulatory alternatives. The managers believed that implementing the regulations most preferred by hunters (as revealed by the survey) would increase waterfowl hunting participation (a possible goal of their management).

Is this conclusion reasonable? While it appears intuitive, it oversimplifies the complexity of attitudes. To illustrate this complexity, consider contemporary conceptual approaches to attitudes. These approaches (e.g., Cohen & Reed, 2006; Petty, Briñol, & DeMarree, 2007) embrace the idea that we can simulta-neously hold both positive and negative attitudes about an issue. These theories suggest that factors, such as context and the person's motivation, dictate whether positive or negative attitudes predominate.

Regardless of its complexity, the attitude construct is used regularly in most applied social science disciplines. Manfredo, Teel, and Bright (2004) reported that attitude studies are the most prevalent type of investigation in human dimensions of natural resources. Why are attitude studies so popular? First, *attitudes offer a parsimonious, easily understandable way to describe a group's thoughts on a specific issue*. Attitudinal responses can be measured by adminis-tering fixed-format response scales and clear survey questions to a sample of people from the population of interest. The distribution of responses can then be summarized using univariate summary statistics. Both managers and stakeholders can accurately interpret these simple results (e.g., "Ninety percent of a stakeholder group supports a proposed management action."). Moreover, citizens of developed nations receive daily exposure to the results of attitude studies (e.g., public polls on political and social issues) through the media.

Second, attitude studies are popular because the information they provide *may help us predict and influence human behavior*. Theory suggests that attitudes are the proximate cause of behavior (Fishbein & Ajzen, 1975). If we know people's attitudes, we can predict how they will respond to new management initiatives, how they will vote, whether they will participate in a particular form of recreation, if they will buy certain products, etc. If we know people's attitudes and understand why they hold these attitudes, *we may be able to influence their behavior*. For example, a person may have a positive attitude toward, and intend to sign, a petition that supports oil drilling in the Artic National Wildlife Refuge in Alaska. He may hold this attitude because he believes that drilling would benefit Alaska's economy. By informing the person that the oil reserves in the Refuge are relatively low, and that the economic benefits would be minor, we could change his attitude and intended behavior.

Historically, research has questioned whether attitude studies have enough validity to support these claims of their utility (e.g., Heberlein, 1973); however, attitude research remains strong, and many researches and practitioners believe that attitudes can be useful in behavioral prediction and behavior change.

Third, attitude studies are popular because of *our own self-awareness of the evaluations we make*. We report our reactions to a topic and give reasons for such reactions. We express our attitudes daily in our behavior, and our attitudes explain our behaviors and the purpose of those behaviors to others. If I am asked why I will engage in a particular behavior, I can give specific explanations; for example, I will fish this weekend because I believe I will catch fish. Because of this self-awareness, attitude studies are relatively easy to conduct. By systematically questioning a sample of people, we can describe their attitudes.

To summarize, attitude studies are popular because they are easy to interpret, they help anticipate or change behavior, and they are relatively easy to conduct.

As a final point, *use of the attitude concept is common because it is central to the development of other, more complex or more topic-specific concepts that describe how people evaluate phenomena*. Allport acknowledged this by saying "the attitude unit has been the primary building stone in the edifice of social psychology" (1954, p. 45). In the applied area of the human dimensions of wildlife (HDW), for example, attitudes help explain customer satisfaction (e.g., most uses of satisfaction are an evaluative measure), conflict (e.g., contradictory evaluations between two or more persons), crowding (e.g., the evaluation of number and type of people in one's immediate surroundings), environmental risk (e.g., evaluations of threats from an occurrence), rapid rural appraisal (e.g., techniques for evaluating the social context to assist conservation or agricultural development efforts in developing countries), and non-market valuation (e.g., people's evaluation of an object such as "scenic vistas" that are not traded in a market place).

Despite the frequent use of the attitude concept, and perhaps because of the ease of conducting studies, generalizations drawn from attitude investigations often reflect a poor understanding of the concept. Misunderstanding of the attitude concept can be found in studies where:

- General attitudes are used to predict specific behaviors.
- Attitudes toward one behavior (e.g., management regulations) are used to predict different behaviors (e.g., hunting participation).
- Publics are asked about highly technical topics of which they have little understanding.
- Questionnaires are developed that influence the attitudes they are trying to assess.

This chapter explores the study and use of the attitude concept in hopes that clarification of the concept will enhance its applications.

Defining Attitudes

At the core of the concept of attitude are people's evaluations of their surroundings. Fazio, Chen, McDonel, and Sherman (1982) defined attitudes as "an association, in memory, of an evaluation with an object" (p. 341). For example, when asked about ice cream (object), most people have a positive evaluation. We would say they have a positive attitude toward ice cream.

While evaluation appears effortless, it involves a series of complex processes, namely receiving and interpreting information, storing information in memory, and retrieving information. These evaluative capabilities emerged in humans because these capabilities led to approach or avoidance behavior that increased survival and reproduction. (Lang, Bradley, & Cuthbert, 1990; Duckworth, Bargh, Garcia, & Chaiken, 2002).

Evaluation occurs incessantly in our daily lives. Among the attitudes formed through the evaluation process, some form quickly and others form slowly. Although a small group of theorists propose that attitudes are constructed each time they are needed (Schwarz & Bohner, 2001), most researchers believe that attitudes can be constructed spontaneously and are also stored in memory and retrieved in situations where they are needed. Of these many attitudes, some remain in memory for a lifetime, some are retained for a short period, and others are forgotten quickly.

While the core of the attitude concept is evaluation, theorists believe that attitudes may involve three different components: an affective component (emotions a person feels toward an attitude object), a cognitive component (beliefs a person holds about an attitude object), and a conative component (behavior related to the attitude object). This approach is called the tripartite definition of attitudes.

An attitude object may have all of these components, or it may just have one or two of these components. For example, a person may form a positive attitude toward donating money to a wildlife habitat fund because that person

- Feels generous about donating to a worthy cause (affective component).
- Believes the money will be used for protecting wildlife habitat and is tax deductible (cognitive component).
- Has donated to similar causes in the past (conative component).

As these cognitive, affective, and behavioral components combine, a person may form a general positive or negative evaluation about donating to a wildlife habitat fund. Once this general attitude is formed, it re-shapes its components to increase the correlation between the attitude and the feelings (evaluative-affective consistency), beliefs (evaluative-cognitive consistency), and behavior (evaluative-behavioral consistency) that created it.

Approaches like the tripartite definition of attitudes propose that the associations formed among how we feel, what we believe, and how we act are highly consistent. More recent approaches suggest that the cognitive, affective, and conative components can be inconsistent and even contradictory. This inconsistency in component evaluations suggests the existence of multiple attitudes toward a single attitude object. This notion is supported in recent advancements that propose a dualistic model of attitude theory (Gawronski & Bodenhausen, 2006; Smith & DeCoster, 2000; Wilson, Lindsey, & Schooler, 2000). Wilson et al. (2000) proposed that humans have both implicit attitudes and explicit attitudes. Implicit attitudes are evaluations that people activate automatically without effortful, thoughtful processes (Greenwald & Banaji, 1995; Wilson et al., 2000). Explicit attitudes are evaluative judgments that are based on the deliberation of, or syllogistic inferences derived from, information relevant to a given situation (Gawronski & Bodenhausen, 2006). Forming explicit attitudes requires conscious cognitive work. We will explore these two types of attitudes later in this chapter.

The majority of research conducted in applied fields measures explicit attitudes. This is due to the fact that the method used to elicit attitudes is intertwined with the nature of the attitude. The next section highlights differences in explicit versus implicit attitude measurement approaches.

How We Measure Attitudes

How we define attitudes influences how we measure them and how the concept is applied to wildlife management issues. Both wildlife professionals and stakeholders accept that attitudinal information contributes to more effective wildlife and habitat management. However, applying theoretical attitudinal concepts requires empirical study and the ability to measure attitude concepts systematically.

How do we measure attitudes? There are two broad approaches (Dovidio, Kawakami, & Gaertner, 2002). The first is *explicit attitude measurement*, which measures evaluations that a person is consciously aware of and can express. The second is *implicit attitude measurement*, which measures evaluations that are automatic and that function without a person's awareness or ability to control them.

Measuring explicit attitudes. Measures of explicit attitudes are, by their very nature, self-reports. Several prominent attitude scaling methods have been developed and used over the past decades. These scales generally require survey respondents to choose among two or more alternative answers to statements that reflect their evaluation of an attitude object or issue. The most common of these scales include *Thurstone's Equal-Appearing Intervals Scale, Likert's Method of Summated Ratings, Guttman's Cumulative Scaling Method,* and *Osgood's Semantic Differential Scale.* For specific descriptions of the development and use of these scales, see Miller (2002) and Oskamp and Schultz (2005).

Measuring implicit attitudes. Measuring implicit attitudes is quite different from measuring explicit attitudes. While a number of approaches have been used to measure implicit attitudes, the most predominant make assumptions about a person's attitude toward an object based on that person's response time to a stimulus; often response times are measured using computer software. The most predominant of these was developed by Greenwald, McGhee, and Schwartz (1998) and is called the Implicit Association Test (IAT). IAT utilizes computer software and people's reaction times to measure the strength of association between category concepts and valuation concepts. For example, consider two descriptive categories, called "hunting activity" and "wildlife disease," and evaluative category of "pleasant" and "unpleasant." In Fig. 4.1, a study participant would see each of the computer screens at separate times. For both computer screens, the participant would be asked to place the descriptor (e.g., muzzle loader) into one of the two groupings on each screen by striking one of two keys on the keyboard as fast as possible.

The assumption is that if participants hold a positive attitude toward hunting, they will place "muzzle loader" into the "hunting or pleasant" category of Computer Screen 1 fastest because that pair is most compatible with their attitudes; they would more slowly categorize "muzzle loader" for Computer Screen 2 because both pairs are incompatible. Conversely, if their attitude toward hunting was negative, the quickest categorization would be for "hunting or unpleasant" on Computer Screen 2.

Researches have questioned the IAT procedure's validity and the correlation between implicit and explicit attitudes' measures. Category ordering and word familiarity are potential validity threats to IAT (Ottaway, Hayden, & Oakes, 2001); however, research has shown that even while controlling for these effects, hypothesized IAT effects still occur (Greenwald, 2004). This suggests that this method of measurement has internal validity (it is capable of measuring the implicit attitude concept).

The correlation between implicit and explicit attitude measures. Researchers have found relatively small correlation between explicit and implicit measures of attitude (e.g., Ottaway, Hayden, & Oakes, 2001, in a study on racial attitudes). If differences are due primarily to measurement error, one can assume that implicit and explicit measures of attitude toward an object are measuring the same concept (Cunningham, Preacher, & Banaji, 2001). The lack of correlation may instead suggest that implicit and explicit attitude measures are measuring separate concepts (Rudman, 2004); however, much of the research on the relationship between implicit and explicit attitudes has addressed *when* implicit and explicit attitudes are correlated. Nosek (2004) suggested that the correlation between the two attitude measures increases when people (a) are unmotivated to present themselves

	Computer Screen 1		Computer Screen 2	
Fig. 4.1 Implicit association test: example computer screens	Hunting or Pleasant	Wildlife Disease or Unpleasant	Hunting or Unpleasant	Wildlife Disease or Pleasant
	Muzzle Loader		Muzzle Loader	

favorably in a given situation, (b) hold important and stable attitudes, and (c) hold attitudes that they believe are different than the average person's.

We know very little about implicit attitudes because they are difficult to measure. However, it is likely that implicit attitudes greatly influence human response to wildlife, particularly if learning quickly about wildlife threats is innate (see Chapter 2). Unfortunately, as of now, there are virtually no studies that explore implicit attitudes toward wildlife; this should be a priority for future investigations.

What Functions Do Attitudes Serve?

As mentioned in the preceding paragraphs, the human capacity to evaluate emerged and was retained because it helped humans adapt to their environments. At a more individual and day-to-day level, people use attitudes for a variety of purposes. Smith, Burner, and White (1956) and Katz (1960) researched and described the motivational basis of attitudes. They identified taxonomies of attitude functions. These taxonomies include *object-appraisal*, *value expressive*, *social adjustment*, and *ego defensive*.

The Object-Appraisal Function

The most recognizable function of attitudes is object appraisal. Attitudes serving this function allow people to classify attitude objects as being either consistent with or inconsistent with their goals; they then respond toward the attitude object in a way that best serves them (Thompson, Kruglanski, & Spiegel, 2000).

Katz (1960) identified two components of the object-appraisal function, the *knowledge component* and the *utilitarian or instrumental component*.

The knowledge component. In the knowledge component, attitudes allow people to organize and simplify their perceptions of a complex environment (Eagly & Chaiken, 1993); this simplified perception serves as a frame of reference for interpreting the world. For example, a positive attitude toward the environment may serve as a framework for interpreting information people receive from the daily news they watch, the policies and political candidates they support, the lifestyle they pursue, their mode of transportation, and their evaluation of other people (i.e., a policy is good or bad because it is pro-environment; a person is good or bad because of his or her attitudes toward the environment). In this way, the attitude helps people filter and classify the barrage of information they receive.

The utilitarian component. The utilitarian component suggests that attitudes provide guidance in maximizing rewards and minimizing punishments. We support things that benefit us and oppose things that harm us. For example, a sheep rancher who believes that the lethal control of coyotes reduces damage to his livestock may develop a positive attitude toward poisoning coyotes because it allows him to maximize self-reward. A politician may adopt a positive attitude toward protecting a specific natural area in order to garner votes from environmental interest groups.

The Value-Expressive Function

Values (discussed in Chapter 7) are broad overarching beliefs that serve as goals across many situations and contexts. They are typically ideals that represent end states (e.g., being free) or desirable ways of behaving (e.g., treat others humanely). According to the value-expressive function, attitudes can express personal values and other core aspects of how individuals view themselves (Maio & Olson, 2000). That is, a person might express an attitude because it is consistent with her "ideal self."

The following example contrasts the value expressive and utilitarian functions of attitudes. Some people support hunting because it is linked to a need to feed their families, an illustration of the utilitarian function of attitudes. Others oppose hunting because they hold a fundamental belief that all creatures, human and otherwise, have rights. This attitude toward hunting is an expression of a deeply held belief and hence takes on a value-expressive function.

The Social-Adjustment Function

The social-adjustment function is similar to the value-expressive function in that it reflects a person's values. However, while the value-expressive function confirms a person's self-concept, the social-adjustment function facilitates relationships with other people and groups (Smith et al., 1956). Expressing attitudes that are pleasing to others or that are consistent with the perceived norms and values of a particular group can assist perceptions of commonality. Such commonality might facilitate short-term cooperation or long-term acceptance by that group or individual. In many situations, we hope to cooperate or communicate with others by voicing an attitude that conveys commonality with them. For example, managers who collaborate with ranchers are typically careful to express attitudes that convey commonality and avoid expressing attitudes that convey difference. The acceptance that results provides a basis for collaboration.

The Ego-Defensive Function

Attitudes that have an ego-defensive function protect the individual's ego or self-esteem. Much of the research on this function of attitudes has been conducted in relation to prejudice against other people or groups. People may unconsciously project their own feelings of inferiority onto another group, thereby bolstering their own egos by feeling superior to members of the out-group (Katz, 1960). These attitudes may not be grounded in realistic percep-tions of the attitude object. For example, avid hunters may perceive that the support of gun control by some groups threatens their right to hunt. As a result, they may adopt the attitude that gun control lobbyists are weak, unpatriotic, and anti-American.

Why people hold the attitudes they do may not always be apparent. Attitudes help us get what we want and avoid what we do not want; they help us organize new information; they help us manage our social interactions and form alliances; they convey what we stand for, and they protect our identity. The attitudes we hold adapt us to a complex world.

Topics in Attitude Theory

Researchers have studied attitudes extensively for decades; they have created many well-founded conceptualizations of attitudes across a variety of contexts. While it is beyond the scope of this chapter to review all conceptualizations of attitudes, this section recognizes some prominent theoretical themes and exemplifies some notable theories.

Attitudes Are Consistent with Other Attitudes and Beliefs

An early tradition in attitude theory explored the association among attitudes. The fundamental notion of these theories was that people strive toward consistency among their attitudes and when inconsistency is made obvious to them, they change.

Balance Theory. A prominent approach here was *Balance Theory* (Heider, 1946). This theory analyzed the knowledge structures among sets of attitudes using the following approach. Balance was examined by identifying three elements of a situation. One element is the *reference person*, a second element is the *other person*, and a third element is the *impersonal entity, or thing*, which can include physical objects, issues, and values (Eagly & Chaiken, 1993). According to *Balance Theory*, a balanced state exists if the relation between each of the three elements (represented as a triad) is either positive in all respects or if two relations are negative and one is positive.

For example, consider the situation where hunting is *the impersonal entity* for which two people hold an attitude. Person A (*the reference person*) is an avid hunter and Person B (*the other person*) is highly involved in the animal rights movement and is opposed to hunting of any kind. A balanced triad would exist if Person A expressed dislike for Person B (two negative relations and one positive relation). If Person A likes Person B, an unbalanced triad exists (two positive relations and one negative relation). In simple terms, Balance Theory states that a balanced state exists when we agree with friends and disagree with enemies on an issue. On the other hand, disagreement with friends and agreement with enemies are unbalanced or unsteady states. An unbalanced state cannot be maintained. Over time, it will transform into a steady state either by a change in Person A's perception of Person B or by a change in person A's perception of the impersonal entity (the issue creating the inconsistency).

Balance Theory was most popular in the 1960s; after that, interest in the theory declined. However, with the increased interest in associative networks, which are drawn from cognitive psychology, Balance Theory's basic tenets reemerged (e.g., Anderson, 1983; Bower, 1981; Judd & Krosnick, 1989).

Associative Network Theory. Associative Network Theory views separate attitude objects as nodes that are linked within an individual's memories, and similar to *Balance Theory*, it addresses inter-attitudinal structure. Judd and Krosnick (1989) characterized attitudes as positive or negative signs attached to an attitude object and suggested that several attitudes can be linked into a network that requires balance or consistency.

For example, an individual's support for strict gun control might have a negative implication for (that is, be perceived as detracting from) one's attitude or support for trophy hunting; this may be further connected to that individual's support for animal rights issues. These linkages in the network not only would be characterized by balanced valences between like and unlike attitude objects but also imply a strength of association, which is reflected in the number of times that a node, or connection between one attitude and another, has been activated in the past (Judd & Krosnick, 1989).

Attitudes Are the Result of Behavior

A theoretical tradition popular in the 1960s and 1970s proposed that people report attitudes that justify or confirm their behavior. These theories noted research showing weak attitude–behavior relationships (Bem, 1972). The effect of behavior on attitudes was originally cast as an argument against the idea that attitudes cause behavior. More recently, researchers accept both of these notions, suggesting past behavior can certainly influence behavior, but does so through first affecting a person's attitude toward a future behavior. The effects of past behavior can be illustrated by considering an individual who has recently toiled planting gardens in her backyard in order to provide a tract of habitat for birds and other small wildlife. As a result of this behavior, and its positive consequences, that person might adopt a positive attitude toward broader ecological restoration of open areas within a large metropolitan area.

Cognitive dissonance. A prominent and classic motivational theory, *cognitive dissonance* (Festinger, 1957) suggests that people's attitude change in response to increased dissonance, or negative feelings, caused by engaging in certain behaviors. Cognitive dissonance arises when people engage in behaviors that are inconsistent with their personal self-concepts, how they want to be viewed by others, or when the behavior has other negative consequences (Petty & Wegener, 1998). The person is motivated to change his or her attitude so it is consistent with the past behavior, thereby reducing the negative feelings, or dissonance, caused by the person's actions.

For example, if a person who believes in and participates in animal rights demonstrations finds himself enjoying a weekend fishing trip with friends, he may experience cognitive dissonance because of his animal rights attitudes. To reduce the dissonance, he may temper his animal rights attitude or adopt the attitude that his animal rights beliefs do not apply to fish.

Attitudes Are the Proximate Cause of Behavior

There is no topic in attitude theory that has provoked more debate or attention than has the validity of the attitude–behavior relationship. It is an enduring question that has spawned the emergence of new theories, provided the basis for rejection of other theories, and stimulated ongoing refinement of prior approaches. At present, there is general agreement that attitudes are the proximate cause of behavior. A select few of the theories that have guided acceptance of this notion are reviewed here.

The Theory of Reasoned Action. One of the most influential approaches to attitude over the past 35 years is Fishbein and Ajzen's Theory of Reasoned Action (TRA) (Fishbein & Ajzen, 1975). It emerged when theorists began to doubt the validity of attitude–behavior relationships. Studies suggesting poor attitude–behavior correlations were the rule (Wicker, 1969).

TRA proposed that behavior is a function of a *behavioral intention*. Behavioral intention is a person's belief about how he or she will behave in a specific situation. Behavioral intention is a function of a person's *attitude* and *subjective norm*. An attitude is an evaluation of a specific behavior. For example, the statement "I am favorable about taking a trip to go fishing this weekend" reveals my positive attitude toward that behavior. Subjective norm refers to the impact of important individuals and groups on a person's behavioral intention, for example, "My friend wants me to go fishing with him this week-end, and I do not want to disappoint him" (see Boxes 4.1 and 4.2 for a description of how TRA can be used).

TRA emerged from a family of models that proposed that a person's behavior is a function of his or her consideration of the outcomes associated with that behavior. Outcomes are evaluated by the likelihood that they will occur (expectancy) and their desirability (valence). As seen in the example provided in Box 4. 1, TRA proposes a mathematical formula that represents how outcomes are evaluated and processed.

Variants of the Theory of Reasoned Action. Ajzen (1985) introduced a modification to TRA which recognizes the distinction between goals people might have (lose weight, stop smoking) and their ability to achieve them. He called this the *Theory of Planned Behavior* (TPB) and expanded TRA to include behaviors that may not be under the volitional control of the individual. This theory has the same components as TRA but adds a factor called *perceived behavioral control*, which is hypothesized to directly influence both behavior and behavior intention. *Perceived behavioral control* is a person's belief about whether he or she has the ability to accomplish a particular behavior. For example, although I have a positive attitude toward going fishing and strong normative pressure encourages me to fish, I may also be concerned that I do not have the equipment and resources to go on a fishing trip and that I am not adequately competent to successfully engage in the activity. This may prevent me from going on the fishing trip despite my positive attitude and normative pressure toward the activity.

Other variables have been suggested as possible omissions from TRA which, if included, would enhance behavior prediction. These includes variables such as moral obligation (Schwartz & Tessler, 1972) and self-identify (Charng, Piliavin, & Callero, 1988).

Warshaw, Shepherd, and Hartwick (1982) and Bagozzi and Warshaw (1990) modified TRA while developing a theory of goal pursuit. The behavior of interest is *trying to achieve a goal*, which is a function of *intention to try*, which, in turn, is a function of *attitude toward and subjective norm regarding trying to achieve a goal*. Attitude toward trying is composed of *attitude toward success* weighted by the *likelihood of success* and *attitude toward failure* weighted by the *likelihood of failure*. An additional *attitude toward the processes involved in trying* is also included.

Triandis (1980) proposed a model similar to TRA but included an *affect* toward the act that included an emotional component that was missing from TRA.

At Times, Attitudes Do Not Affect Behavior

With gradual acceptance of the notion that attitudes do, indeed, cause behaviors, researchers began to look more carefully at the instances where this does not occur. Fazio (1995), a pioneer in this area, suggested that the accessibility of an attitude is a critical determinant of whether the attitude will affect behaviors. Fazio (1995) proposed an attitude–behavior model that suggests that people may not always think deliberatively about an attitude object when coming into contact with and engaging in behavior toward it. This model emphasizes the role of motivations (M) and opportunity (O) as determinants (DE) of how attitudes influence behavior (called the *MODE model*).

When an individual is highly motivated to think deliberatively about an attitude and the relevant behavior and has the opportunity to do so, the attitude affects behavior in the way that models like TRA suggest (Sanbonmatsu & Fazio, 1990). However, there are times when this deliberation does not occur. In these situations, deliberative models like TRA do not predict behavior. In cases such as these, only attitudes that are automatically activated from memory (without deliberation) will guide behavior (Fazio, 1995). The extent to which an attitude can be activated from memory is called *attitude accessibility*.

What makes attitudes accessible? According to Fazio (1995), attitudes formed through direct experience with an attitude object are more accessible than attitudes formed through more indirect means, such as reading a book or watching a TV show. In addition, attitudes repeatedly supported by an individual's past behavior are more likely to be automatically activated in future situations. Furthermore, an attitude that is automatically accessed without conscious thought results in a biased perception of the object in the immediate situation, and behavior simply follows from these [biased] perceptions without any deliberation.

For example, consider a proposal to increase the harvest of timber in a forest located near an economically depressed region. The region's economy depends heavily on the timber industry, and it has been determined that increasing timber harvest will have little or no negative environmental impacts. An individual who has historically supported the protection of natural resources over their economic use may automatically access a deeply held protectionist attitude and vote against increasing timber harvest without considering the benefits of increasing the harvest in this particular region.

Why is accessibility important? An attitude must be accessible in order to influence behavior. If a stable attitude is inaccessible, a person will usually form an attitude, but that attitude might be influenced by context-specific factors and not predict later behavior. Manfredo, Yuan, and McGuire (1992) illustrated this by showing that people with more experience in Yellowstone and people more involved in discussions about fire had higher attitude–behavior correspondence on the issue of controlled burn fire policy.

Accessibility bears an important implication: researchers must be cautious when asking respondents about complex or obtuse topics or issues toward which respondents are unlikely to have attitudes. To illustrate, state fish and wildlife agencies often want to obtain answers from the general public to questions such as "How would you rate the performance of the state fish and wildlife agency?" Because a high percentage of the general public cannot even correctly identify which agency in a state manages wildlife, it is unlikely they have an attitude toward that agency's performance. Managers in the United States often want responses to detailed questions that require a high level of knowledge, often beyond that of the public. While the public may give answers to such questions, their durability and consistency over time will be low.

There Are Two Types of Attitudes, and They Can Be Contradictory

Fazio's work and dual process models of persuasion (Eagly & Chaiken, 1993; Petty & Cacioppo, 1986) in the 1980s encouraged researchers to consider the possibility that humans engage in evaluation in more than one simple way. This research has explored the notion that people may hold two attitudes toward an object and that these attitudes result from different mental processes. Attitude change research has created interest in this topic.

To illustrate, researchers assumed that the new attitude replaces the old attitude in the individual's mind (Wilson et al., 2000). However, many theorists now suggest that, while the new attitude is stored in memory, the original attitude is not replaced; it remains in memory. This notion is reflected in current models of attitudes that propose a dualistic approach (Gawronski & Bodenhausen, 2006; Smith & DeCoster, 2000; Wilson et al., 2000).

Wilson et al. (2000) proposed the existence of both "implicit" attitudes and "explicit" attitudes. As noted earlier in the chapter, an implicit attitude occurs automatically and is simply present in memory; the person experiencing the implicit attitude has little conscious awareness of how the attitude emerged (Greenwald & Banaji, 1995; Wilson et al., 2000). Explicit attitudes are evaluative judgments that the individual consciously creates by deliberating situation-relevant information (Gawronski & Bodenhausen, 2006).

Explicit and implicit attitudes toward an object can coexist in memory, and they may be inconsistent; therefore, even when explicit attitudes are activated, the implicit attitude can influence a person's response toward an attitude object. Because of the interaction among explicit and implicit attitudes, the response the individual expresses may differ from the response that would be elicited by an explicit attitude alone (Wilson et al., 2000).

These two types of attitudes result from different mental processes. Smith and DeCoster (2000) proposed that implicit attitudes arise through associative processes of learning and memory. Associative learning occurs through repeated pairing of an object and an evaluation (Smith & DeCoster, 2000). Because implicit attitudes occur through repeated pairing, they take a long time to form. Once formed, however, they are enduring. A person does not actively engage in the retrieval of such an association; it occurs quickly, automatically, and effortlessly. For example, the repeated paring of a particular place and positive social interactions with a parent might create an implicit positive attitude toward the place, which is automatically retrieved when the location's name is mentioned or when the individual glances at a photo of the location.

Explicit attitudes arise through active cognitive engagement that has been described as rule-based (Smith & DeCoster, 2000), or propositional processing (Gawronski & Bodenhausen, 2006). Smith and DeCoster (2000) contended that, in rule-based processing, individuals derive evaluations by processing available information. Rule-based processing is structured by language and other symbolically referenced rules (i.e., occurs by using language). If motivated and capable, people can form explicit attitudes after just one experience.

Smith and DeCoster (2000) suggested that that these two human memory systems (implicit and explicit) evolved because they helped humans meet two competing survival demands. Associative learning (implicit) creates long-term, stable knowledge that helps humans understand their typical environmental conditions, while explicit learning allows humans to learn rapidly in novel situations. The formation of explicit attitudes allows for quick adaptation of humans to their environment.

Extending the dual attitudes perspective, Cohen and Reed (2006) proposed a *multiple pathway anchoring and adjustment model* (MPAA) that combines the traditional anchoring and adjustment view of attitude formation with a dual attitudes perspective. In their model, when a person is exposed to an attitude object in a given context, that person either uses an "outside-in" (explicit attitude processes) or an "inside-out" (implicit attitude processes) mechanism to form and store an attitude. In this model, two attitudes of opposite direction toward the same object may coexist. When a person is exposed to a previously assessed attitude object, the previously formed attitude may be accessed. If the previously formed attitude is not accessed, the individual must construct an attitude from available contextual information. At this point, either the previously formed attitude or the recently constructed attitude is checked for *representational sufficiency*; that is, is it representative of the situation? If not, then new information is retrieved and a new attitude is constructed that is sufficient. The attitude is then assessed for *functional sufficiency* for guiding behavior; if not, the attitude is adjusted until it can guide behavior.

This model might have important uses in the examination of human-wildlife relationships. If people have well-established, implicit attitudes toward wildlife (enhanced through genetically prepared learning), this may guide their evaluation of wildlife-associated issues. People will only use effortful cognition if their implicit attitudes cannot guide their responses to a situation. For example, people may have negative implicit responses to threatening species, but they may judge these implicit responses as inadequate when evaluating questions about reintroducing species. When implicit responses are judged inadequate, a person deliberates. If we can understand what motivates deliberation, we might derive important findings.

Conclusion. To summarize, researchers have developed theory that examines various aspects of the attitude concept. Theory has helped us understand the consistency of our attitudes; we accept that attitudes cause behavior and that attitudes are formed, in part, by our behaviors. Our attitudes do not always predict our behavior. By exploring these situations, theory recognizes that people may hold two attitudes toward the same object and that two separate mental processes may produce these attitudes. In the next section, we examine attitude theories that have been applied in the area of human dimensions of wildlife.

Attitude Theory Applied in Human Dimensions of Wildlife

Few theoretical approaches to attitudes have been applied to human dimensions of natural resources. The descriptive, non-theoretical approach to attitudes is the most popular approach.

Descriptive Approaches

Popular among the management community is polling, with the use of single-item attitude measures that report on public reaction to current issues. Due to the ease of conducting and interpreting these studies, they have become quite common among fish and wildlife agencies in the United States, and private polling firms have emerged to serve that need (e.g., Duda, Bissell, & Young, 1998). These studies provide timely information to managers, but their generalizability is typically quite restricted and, in cases, their validity questionable. The generalizations from descriptive studies are typically restricted to the place, time, and specific population studied. The ability of these approaches to explain why attitudes are held is also descriptive and correlational. They might, for example, explore the association of a particular issue with available descriptive variables (e.g., do results vary by high- versus low-income residents, rural versus urban residents, males versus females?).

Identifying the "dimensions of attitudes" toward a particular topic is another popular approach. This is similar to the approach used to develop value scales (value scale construction is described in Chapter 6). Multiple survey items are developed on topics that are related to the topic of interest. The items are administered and factor-analyzed with the resultant item; groupings (i.e., the factors) represent the attitudinal dimension. While this approach is probably useful in exploring basic and enduring patterns of thought among people (i.e., their values), it inadequately captures the processes by which people attend to information or retrieve information in forming an attitude. For example, much of the information represented in items on this type of survey is unlikely to occur to the person without the prompting, and the instrument itself actually influences the attitude that is reported. Hence, findings from such a survey may be inaccurate in predicting the behavioral response of the population of interest; i.e., the study sample results were shaped by item prompting and, as a result, are dissimilar from the unprompted attitudes of the population. This would be a problem when studying attitudes on topics where people have little information or experience, a common situation for wildlife agencies conducting surveys of the general public.

Theoretical Approaches

Beyond these descriptive studies, there are several cases where a theoretical approach has been introduced in the human dimensions of wildlife or natural resources area, but the application remains a somewhat isolated case. For example, Stewart (1992) used cognitive dissonance theory to explain why hikers change their ratings of trip motivations so the ratings are consistent with the experience that occurred; i.e., hikers who felt they did not get a good workout, lowered the importance of physical exercise from their a priori rating, as a motivation for going hiking (reducing dissonance).

Jackson, White, and Schmierer (1996) used attribution theory to understand how people develop evaluations of their tourist experiences. People were more likely to attribute positive experiences with internal attributions (i.e., positive experiences occurred because of something the tourists made happen) and negative experiences with external attributions (negative experiences were caused by occurrences beyond their control).

Bright (1997) applied theory about attitude strength when examining reactions to recreation management strategies on the Arapahoe and Roosevelt National Forests and Pawnee National Grassland. Extreme attitudes were significantly better predictors of voting behavior than moderate attitudes. Additionally, attitudes were stronger when people had high certainty rather than when they had low certainty and when the issue was of high personal relevance rather than low personal relevance. Similarly, Manfredo, Yuan, and McGuire (1992), testing conclusions drawn from Fazio's attitude accessibility theory, found prediction of support for controlled burn fire policies from attitudes increases as does one's direct experience with fires and the extent to which one has engaged in repeated conversation about the topic.

Stated Choice Models

Beyond these more isolated applications of attitude theory, a recent trend in the recreation and the willingness-to-pay literature has been to apply stated choice models of preferences (Haider, 2002; Louviere, Hensher, & Swait, 2000). These models focus on data collection and analysis methods instead of attitude theory development, but their popularity merits mention here. Conceptually, these approaches employ a rationalist approach to attitudes. That is, it is assumed that people will make choices by evaluating the attributes of the alternatives presented to them. A person will choose the alternative that maximizes his or her own benefits. Methodologically, stated choice models present people with an array of behavioral choices. The choices are carefully developed so that the researcher can infer from the participant's choices which attributes were most important. When there are clear policy alternatives with definable attributes of choice, this approach is advantageous; however, one of the major weaknesses of this approach is that it assumes humans maximize utility when deciding among alternatives. A significant amount of research suggests this does not occur (e.g., Tversky & Kahneman, 1974). People mix thoughtful analysis with simple decision rules that rarely reflect a maximization process. So, while the stated choice models are quite useful in prediction of choice, they are weak in reflecting the evaluation processes that guide behavior.

The Theory of Reasoned Action and the Theory of Planned Behavior

Aside from these applications of attitude theory, the most frequently applied attitude theories in the area of human dimensions of wildlife are Fishbein and Ajzen's Theory of Reasoned Action (TRA) and Ajzen's revision of TRA, Theory of Planned Behavior (TPB). TRA is frequently used because (a) it offers a parsimonious explanation of the structure and influence of attitudes, (b) its methods are described clearly, and (c) it produces results that can be readily interpreted with practical implications (Ajzen & Fishbein, 1980). Moreover, this thoroughly tested theory has strong predictive validity, which is a critical criterion in applied fields (Sheppard, Hartwick, & Warshaw, 1988; Albarraćin, Johnson, Fishbein, & Muellereile, 2001; Armitage & Conner, 2001).

TRA/TPB focuses on explicit attitudes, attitudes that are the result of conscious, deliberative thought. For readers interested in more detail about TRA and TPB, Boxes 4.1 and 4.2 provide two different illustrations of the use of TRA/TPB in human dimensions of wildlife. While weaknesses of the TRA have been noted (see, for example, Eagly & Chaiken, 1993), the theory predominates in applied fields and has endured for four decades. Because of its utility, it is likely to subsist.

Factors Affecting Attitude–Behavior Relationships

As noted in the introduction to this chapter, the ease of conducting attitude studies contributes to a high degree of variability in the quality of investigations. While people will, in most cases, answer questions they are asked, the durability and accuracy of their responses vary. Accuracy and durability are a concern when people are asked about (a) highly technical topics that many of them may not understand (e.g., ecosystems, biodiversity), (b) situations for which they have little information (e.g., how a division of an agency is performing, the distribution of noxious weeds in the state), and (c) situations for which they have no experience (e.g., evaluating the performance of law enforcement officers when they have had no contact with them).

Respondents may give answers to questions about these topics, but these responses are unlikely to predict behavior and can readily change over time or with just small amounts of new information. Unfortunately, wildlife managers often rely on attitude studies to provide public opinion on issues that fit these characterizations (technical terms, little information, and no direct experience). Attention to just a few key factors would help investigators avoid these types of problems.

In this section, we examine these key factors and their effect on attitude–behavior relationships. Our examination of these topics is consistent with a shift in concern among social psychologists; that is, research on the attitude–behavior relationship has evolved from examining *whether* attitudes predict behavior to *when* attitudes predict behavior (e.g., Wallace, Paulson, Lord, & Bond, 2005; Glasman & Albarraćin, 2006). In this section, we examine topics that influence the attitude–behavior relationship and help shape our expectations of when strong relationships should occur. Overall, we reiterate our suggestion that future research should expand beyond measurement of explicit attitudes to examine implicit attitudes toward wildlife.

Specificity

While explaining studies that found poor attitude–behavior relationships, Fishbein and Ajzen (1975) introduced a criterion for guiding future studies that requires *attention to specificity*. The essence of this principle is that *attitudes will not predict behavior unless they are measured with corresponding levels of specificity*. Attitude specificity is likely to have been one of the most significant refinements in improving applicability of the attitude concept. Specificity suggests that for attitudes to predict behavior, their focus must match. General attitudes (toward hunting participation) will not predict specific behaviors (taking a trip this weekend to hunt deer in Colorado's Poudre Canyon). Attitudes about objects (e.g., wilderness) will not predict behaviors (visits to wilderness).

For attitudes to predict behavior, the attitude and the behavior must correspond on four levels of specificity: action, target, context, and time. For example, assume we have these four different attitude objects:

1. Taking a recreation trip (action).
2. Taking a recreation trip to view wildlife (action and target).
3. Taking a recreation trip to view wildlife this weekend (action, target, and time).
4. Taking a recreation trip to view wildlife this weekend at Potter Swamp (action, target, time, and context).

We should expect the behavioral correlates of these four different situations to be quite different. For example, questions about attitude objects 1 and 2 are quite general and would be expected to predict general intentions to participate in the future or indices of participation in the past. Questions about attitude objects 3 and 4 would be quite specific and should predict the behaviors described in the statements. However, questions about 1 and 2 would be very poor predictors of the behaviors described in attitude objects 3 and 4. The first and most important step of any attitude study is to clarify the attitude object and the behaviors that this attitude should predict.

Salience

Salience describes the prominence of certain beliefs that comprise a person's attitudes and the extent to which these beliefs routinely occur to an individual in a given situation. To illustrate the importance of this concept, consider the methodological weaknesses of some attitude surveys.

In many attitudinal studies, researchers include on a survey a list of statements that represent beliefs a person might hold about a particular attitude object and ask a sample of participants their level of agreement or disagreement with the statements. Participants' response patterns to these statements is assumed to reveal their attitudes toward the attitude object. This approach ignores salience. The problem is simple: if the researcher asks a person whether she believes a statement, the person provides an agree–disagree response. The survey response does not reveal the reason a person holds the position; the respondent was essentially forced by the researcher's prompting to give a response, regardless of whether she cares about or knows anything about the topic. *Salient* beliefs are those that will most likely come to an individual's mind in a given situation without prompting by a researcher and will more accurately reflect or contribute to a person's attitude toward an attitude object.

Attitude Strength

One of the primary ways by which we contrast attitudes and their ability to predict behavior is by characteristics related to their strength. That attitudes can vary in their strength is obvious as we consider our own preferences over a number of topics. On some topics, we have firm, unfaltering, and resistant-to-change positions and on others we are only weakly committed. Strength is quite important because it can affect things such as the intensity of behavioral response, the consistency over time of one's action, and the likelihood that a person's attitude can be changed.

Some characteristics of strong attitudes include the following:

1. *Strong attitudes are stable.* They remain relatively unchanged over time and would be consistent regardless of context. Most people who are hunters in the United States learned to hunt at an early age and were introduced by a parent. Their attitude toward the acceptability of hunting as a form of recreation tends to remain fairly stable over time and would be quite difficult to change.
2. *Strong attitudes influence how people process and evaluate information.* People tend to process information in a way that is consistent with their existing attitudes. Strong attitudes are more influential in this regard. Even though there may be balanced information in a message, people will focus on arguments that are consistent with their existing attitude. For example, someone with strong attitudes in support of drilling for oil in the Arctic National Wildlife Refuge may process information about the issue in a biased manner such that she automatically rejects the credibility of information that opposes the need for drilling.
3. *Strong attitudes are resistant to attempts at persuasion or attitude change.* Given their stability over time and the influence they have on processing information, strong attitudes are enduring.

4. *Strong attitudes guide behavior*. Attitudes that are more stable, resistant to change, and influential in processing information are also more likely to have an effect on people's behavior.

For overviews of this area, see Cooke and Sheeran, 2004; Petty and Krosnick, 1995; and Glasman and Albarraćin, 2006.

The literature has explored ways to characterize attitude strength. Characteristics include, but are not limited to, extremity, ambivalence, certainty, centrality, and knowledgeability. These are summarized below.

Attitude-extremity. *Attitude-extremity* is *how* favorably or unfavorably a person evaluates an attitude object (Krosnick & Petty, 1995). While two people may agree that shooting coyotes from a helicopter is a good way to control the population, one person may feel that this activity is an excellent method of predator control while another person may feel it is acceptable, but there are other ways with higher acceptability. Extreme attitudes are more likely to predict a person's behavior than moderate attitudes.

Attitude-ambivalence. We are often *ambivalent* in the attitudes we hold. Attitude-ambivalence is the degree of conflict between a person's positive and negative evaluative components of a single attitude object. Attitudes low in ambivalence involve either mostly positive or mostly negative beliefs about an attitude object; attitudes high in ambivalence reflect both positive and negative beliefs about the object.

For example, an ambivalent individual may believe that reintroducing grizzly bears into an area is a good strategy because she strongly believes that these bears are (a) an important part of the natural ecosystem and (b) attractive and interesting animals; this person may also believe (c) it is very likely that the bears will kill livestock, resulting in bears being killed by authorities, and (d) be a potential threat to humans. Items (a)–(d) describe an ambivalent attitude toward reintroducing grizzly bears. Predicting an individual's behavior is difficult when his or her attitude toward that behavior is ambivalent (Conner et al., 2002).

Attitude-certainty. In contrast to ambivalence, a person may hold an attitude with high or low certainty. People are motivated to hold "correct" attitudes (Petty & Cacioppo, 1986). However, people are more confident in the correctness of some attitudes than others. *Attitude-certainty* is defined as a "subjective sense of conviction or validity about one's attitude or opinion" (Gross, Holtz, & Miller, 1995, p. 215).

For example, a devout hunter may hold a strong positive attitude toward firearms used in hunting. He may also, when asked, express a negative attitude toward federal gun control laws. However, it is possible that he is less sure about the correctness of his negative attitude toward gun control because it raises questions of public safety and protection in his mind. Attitudes that people hold with high levels of certainty tend to be stable over time, difficult to change, and predictive of behavior. The level of certainty with which people hold an attitude may also impact the extent to which they processes different types of information.

Attitude-centrality. A certain attitude might also be a central one. The *centrality* of an attitude implies that some attitudes are embedded within a broader network of beliefs a person holds and may be highly influential in that network. Much of the discussion of attitude-centrality relates to the connection between higher-order attitudes (i.e., attitudes toward a specific object or behavior) and more general attitudes or values.

For example, a positive attitude toward drilling for oil in the Arctic National Wildlife Refuge may result from a more general, and central, positive attitude toward humans' right and ability to use nature for their benefit. This central attitude toward human dominion over nature is likely to be linked to other general attitudes (e.g., the prevalence of economic needs over environmental needs) and more specific attitudes (e.g., positive attitude toward commercial development into open space in one's community). As a result of this linkage with other attitudes, any attitude change or persuasion that does occur will occur slowly (McGuire & McGuire, 1991).

Working knowledge. A final characterization of attitudes includes the amount of knowledge associated with it, i.e., working knowledge related to an attitude. Working knowledge represents the information an individual has at his or her disposal when evaluating or processing information about the attitude object (Wood, Rhodes, & Biek, 1995). An individual who understands (a) the history and reasons for the passage of the Endangered Species Act (ESA), (b) the species of flora and fauna that are listed as threatened or endangered, and (c) the arguments for and against endangered species use in several areas of the country is likely to have much stronger attitudes toward the ESA than people who know little about it. High working knowledge about an attitude or attitude object is related to greater prediction of behavior and resistance to persuasion attempts (Woods, Rhodes, & Biek, 1995).

Conclusion

When attitude studies ignore considerations of specificity, salience, and attitude strength, they take the risk that study results will have low predictive validity. Techniques that explicitly account for these concerns are Ajzen and Fishbein's (1980) suggestion for assessing modal salient beliefs, the inclusion of questioning that allows measurement (e.g., certainty), and conducting analysis that examines extremity directly (e.g., Bright & Manfredo, 1995). Using these methods improves the applicability of attitude studies.

Summary

Despite their potential limitations, attitude studies have been the most frequently used investigations in human dimensions of wildlife and natural resources. The concepts of attitude theory (although not necessarily the methods) are relevant whether one is studying illiterate rural residents in developing countries or highly educated residents of developed countries. Researchers will continue to use attitude studies. In these uses, stronger linkages to theory will improve and extend the application and validity of these studies.

In the next chapter, we explore values. While attitudes can vary from situation to situation, values are unchanging. From values arise a consistency in the pattern and direction of attitudes held by an individual; from values, attitudes and norms arise.

Management Implications

It would be difficult to be comprehensive in describing managerial applications of the attitude topic in human dimensions of natural resources. When managers in the United States think of involving a human dimensions perspective, they typically describe their need as an opinion or attitude study. Attitudes not only are conducted in many situations but also are frequently part of other topics studied in human dimensions of wildlife.

Attitude studies are commonly used when managers are interested in representing the interests of publics in the decisions they make, predicting how the publics will behave in certain situations, and anticipating the impacts of various types of alternatives. A few of the applications are described below.

Evaluating Management Alternatives. Managers are often interested in knowing how the public evaluates potential management actions. Do they support or oppose reintroduction of endangered species? Would they support a particular mechanism for raising funds for wildlife management? What are preferred modes of dealing with wildlife involved in human–wildlife conflict? Ironically, one of the arguments against such studies has been that management should not be dictated by public preference. That is generally true. There are many types of information that must be considered in selecting management action and none alone should be the only consideration. That does not mean, however, that public opinion information is not useful or should not be considered. How might these attitude studies be useful? They might reveal findings that suggest ways to effectively communicate a management decision to publics, they might show areas where the public holds inaccurate beliefs, they might show areas of consensus among conflicting groups, they might reconfirm or correct managerial view of public preference, or, in some cases, they may actually suggest that following public preference is clearly the best alternative.

Evaluating Site Conditions or Recreational Experiences. This is a rather fundamental aspect of understanding the success of management programs. Are people satisfied with their fishing or hunting experiences? How would one rate the facilities or educational materials that were available? Did the number of other people present create perceptions of crowding?

Predicting People's Behavior. There are some cases where managers are interested in knowing how stakeholders will behave in a given circumstance. Prominent examples include (1) studies to predict how people will vote on ballot initiatives such as those to ban trapping or hunting or to accept tax increases that will fund wildlife management or (2) studies that attempt to predict recreational participation with regulations changes or with the creation of new facilities. These are the types of studies where predictive validity is a significant concern. It would be important, for example, to know whether a given ballot initiative will pass or not. If provided prior to an initiative being placed on a ballot, it can be important information to prompt compromise solutions. Prediction of recreation behavior is requisite to estimating their economic impacts, i.e., if participation decreases, expenditures and license sales also decrease.

Understanding Current Beliefs as a Basis for Effective Education. Attitude studies are sometimes used to determine people's evaluation of certain topics and the beliefs that support that position. Knowing what people currently believe and what they are interested in learning about can be quite useful in focusing educational efforts.

Facilitating Marketing Efforts. Agencies and NGOs involved in conservation issues are frequently interested in marketing their programs to potential stake-holders. For example, NGOs are frequently involved in securing grassroots funding support. Agencies are increasingly concerned with recruiting people to engage in activities like hunting and fishing. Such marketing efforts can be improved by targeting them toward people with positive attitudes or by presenting promotional materials that focus on aspects of the program that people evaluate positively.

A quick review of articles in journals such as *Human Dimensions of Wildlife* will provide a wide array of additional studies with managerial implications.

Summary Points About Attitudes

- Attitude studies are the most frequent type of study in human dimensions of wildlife and natural resources because

  - They are relatively easy to conduct and interpret.
  - They are easy for study participants to engage in.
  - They offer the promise of behavioral prediction and behavior change.
  - Attitudes are a critical component of many more topic-specific or complex concepts.

- At its core, an attitude is an evaluation of an object. It is interwoven with three components: beliefs about the object, affect or feelings, and behavior. We have two types of attitudes: explicit, which are linked to active cognitive processing, and implicit attitudes, which are built through repeated associations and ultimately become automatically and quickly retrieved.
- Attitudes are measured in many ways. Explicit approaches often use interview or survey responses to fixed-format questions to measure attitudes. The

measurement of implicit attitudes is more difficult because implicit attitudes do not involve cognitive processing.

- Why a person holds a particular attitude might not be readily apparent. They may hold it for

  ○ A utilitarian purpose (liking things that would provide a person positive outcomes)
  ○ Value expressive purposes (to support deeply held beliefs about desired goals or modes of conduct)
  ○ Social adjustment reasons (to gain acceptance by others)
  ○ Ego defensive reasons (poorly founded denigration of others who oppose what a person believes in)

- The development of attitude theory over the past 60 years has focused on the following:

  ○ The consistency among attitudes
  ○ The effect of past behavior on attitudes
  ○ The attitude–behavior relationship, which focuses on factors that influence why, at times, attitudes do not predict behavior
  ○ Recognition that people can hold two contradictory attitudes toward an object

- In human dimensions of wildlife specifically, the preponderance of attitude studies is descriptive and non-theoretical. There are isolated applications of various attitude theories, but the most frequently used attitude theory is Fishbein and Ajzen's Theory of Reasoned Action and Ajzen's Theory of Planned Behavior. These theories have strong predictive validity and yield readily useable results for management and policy decisions.

- Attitude studies frequently conducted attempt to elicit response to topics poorly understood, for which people have no knowledge and no experience. Such studies result in poor predictive validity. Considerations of specificity, salience, and attitude strength will improve the results of these studies.

Box 4.1 Example Application of Fishbein and Ajzen's Theory of Reasoned Action on the Topic of Wolf Reintroduction in Colorado

In the mid-1990s, federal officials were considering whether or not to include Colorado in the Northern Rocky Mountain Wolf Recovery Plan. To assess public reaction to that proposal, a study was conducted to examine the general public's attitudes toward wolf reintroduction. The Theory of Reasoned Action provided conceptual and measurement guidance to the study, and we reproduce aspects of the study here to illustrate the TRA approach.

TRA proposes behavioral intention (BI) is a function of attitudes (ATT) and subjective norms (SN). TPB adds to TRA the notion of perceived behavioral control (PBC), reflecting the idea that your perceptions of constraints will affect your behavioral intention over and above ATT and SN. Both theories suggest that beliefs (cognitions) form the basis of ATT, SN, and PBC. For example, attitudes are represented as

$$\text{ATT} = \sum_{i=1}^{n} b_i e_i,$$

where b is strength of the belief, e is the evaluation of that outcome described by the belief, and n is the number of salient beliefs in the set.

Several characteristics of this formulation are worth noting. First, attitudes are a function of the n beliefs that are salient to an individual, i.e., an important variable examining the attitudes among people is their subjective knowledge regarding the attitude. Second, two qualities about each belief are theorized to be important and proposed for measurement. One is the extent to which a person perceives the belief is likely or true (b) and the other is the positive or negative evaluation that is associated with the belief (e).

Our study of wolf reintroduction in Colorado illustrates the results using this approach (Pate, Manfredo, Bright & Tischbein, 1994). This study showed that, overall, 70.8% of Coloradoans would vote "yes" to reintroduce wolves in the state (a behavioral intention). The measures help us understand why people would vote that way.

The 12 beliefs listed in Table 4.1 were determined, using open-ended elicitation procedures, to be salient for the group being studied. (See Ajzen & Fishbein, 1980, Appendix A, for instructions on conducting an elicitation study.) That is, these are the beliefs that most frequently came to mind for Coloradoans when they were asked about the outcome of their voting to support wolf reintroduction in the state, i.e., the modal salient beliefs. Belief scores are listed separately for those with positive attitudes and those with negative attitudes. Three separate scores are given for each group: the belief

Table 4.1 Beliefs about the outcomes of wolf introduction, from a mail survey of Colorado residents conducted during summer, 1994 (Reproduced with kind permission from the Wildlife Society Bulletin)

Outcomes	Negative $N = 210$		Positive $N = 502$		F
	x	SD	x	SD	
Reintroducing wolves would ...					
...result in large number of wolf attacks on the livestock					
BE Product	−2.59	4.78	2.31	3.47	230.0**
Bad–Good	−2.47	0.99	−1.89	1.15	40.5**
Disagree–Agree	0.86	1.69	−1.21	1.47	263.7**
...result in ranchers losing money					
BE Product	−2.90	4.48	0.83	3.43	142.3**
Bad–Good	−2.24	1.08	−1.62	1.29	37.0**
Disagree–Agree	1.03	1.56	−0.73	1.58	182.9**
...keep deer and elk populations in balance					
BE Product	0.14	3.42	4.41	3.43	225.1**
Bad–Good	1.15	1.40	2.15	0.98	117.6**
Disagree–Agree	0.04	1.71	1.81	1.11	265.7**
...increase tourism in Colorado					
BE Product	−0.89	4.19	0.85	2.90	39.7**
Bad–Good	0.73	1.69	1.02	1.62	4.5
Disagree–Agree	−1.34	1.49	0.30	1.50	176.3**
...result in wolf attacks on human					
BE Product	0.43	4.66	3.88	4.47	84.6**
Bad–Good	−2.48	1.15	−2.32	1.16	2.7
Disagree–Agree	−0.23	1.69	−1.67	1.53	123.0**
....preserve the wolf as a wildlife species					
BE Product	0.71	3.42	5.25	3.44	255.4**
Bad–Good	0.41	1.62	2.39	0.88	437.3**
Disagree–Agree	0.26	1.67	2.00	1.13	255.3**
...return the natural environment back to the way it once was					
BE Product	0.34	3.56	2.99	3.90	70.4**
Bad–Good	0.21	1.47	1.97	1.16	284.3**
Disagree–Agree	−0.95	1.74	0.95	1.75	171.2**
...help people understand the importance of wilderness					
BE Product	−1.96	3.40	3.66	3.74	348.6**
Bad–Good	1.40	1.26	2.49	0.81	187.3**
Disagree–Agree	−1.23	1.59	1.26	1.39	482.4**
...result in wolves wandering into residential areas					
BE Product	−2.06	4.32	1.76	3.70	140.4**
Bad–Good	−2.32	1.05	−1.77	1.18	34.4**
Disagree–Agree	0.77	1.56	−0.84	1.62	148.2**
...result in ranchers killing wolves					
BE Product	0.42	4.53	−3.11	3.53	121.3**
Bad–Good	−0.12	1.75	−1.71	1.22	187.3**
Disagree–Agree	1.80	1.29	1.80	1.10	0.0

Table 4.1 (continued)

Outcomes	Negative $N = 210$		Positive $N = 502$		
	x	SD	x	SD	F
...lead to large losses in deer and elk populations					
BE Product	−0.53	4.26	2.01	3.42	69.4**
Bad–Good	−1.66	1.51	−1.25	1.45	10.9**
Disagree–Agree	0.19	1.80	−1.33	1.39	145.8**
...lead to greater control of rodent populations					
BE Product	0.86	3.69	3.66	3.93	66.7**
Bad–Good	1.49	1.44	1.95	1.22	18.5**
Disagree–Agree	0.53	1.57	1.54	1.38	71.5**

$*P < 0.01$; $**P < 0.001$.
Belief evaluation (BE) product is the multiplication of the agree–disagree scale and the good–bad scale (each ranged from −3 to +3). It ranged from −9 to +9. Note: a positive score could be the result of a negative agree–disagree score (including unlikely the item would occur) and a negative good–bad score (indicating bad).
Scale points included −3 (extremely bad), −2 (moderately bad), −1 (slightly bad), 0 (neither), 1 (slightly good), 2 (moderately good), 3 (extremely good); −3 (strongly disagree), −2 (moderately disagree), −1 (slightly disagree), 0 (neither), 1 (slightly agree), 2 (moderately agree), and 3 (strongly agree)
Source: Pate, J., M. J. Manfredo, A. D. Bright, and G. Tischbein. 1996. Coloradan's attitude toward reintroducing the gray wolf into Colorado. *Wildlife Society Bulletin*, 24(3), 421–428

agreement (agreement that the outcome would occur) score, the evaluation score, and the belief agreement time evaluation score (*be*).

The *be* scoring attains an important meaning in attaining an overall expectancy valence attitude score. It is important to recognize that scoring is always on a +3 to −3 scale for both belief strength (*b*) and evaluation (*e*). Two positives or two negatives multiplied together result in a positive score and contribute to a positive attitude.

For example, look at the *b*, *e*, and *be* mean scores for those with a positive attitude toward wolf reintroduction on the outcome "ranchers losing money." Note that these mean scores indicate that those with a positive attitude believe this would be a bad outcome; however, they disagree that this would occur. A bad outcome that is unlikely to occur (two negatives) contributes to a positive attitude toward wolf reintroduction. They also feel that wolf reintroduction will keep deer and elk populations in check, a positive and likely occurrence contributing to a positive attitude.

In contrast, a positive likelihood and a negative outcome contribute to a negative attitude, as does an unlikely outcome that is positive. For example, those with a negative attitude felt it was likely that wolves would wander into residential areas and that was a negative outcome (contributing to a negative attitude). Compare that to scores for those with a positive attitude who thought that wolves wandering into residential areas would be negative; however, they also felt that was unlikely (contributing to a positive attitude).

Overall, by examining these results, we are able to understand not only people's evaluation of a policy toward wolf reintroduction but also the network of beliefs that serve as a basis for support and opposition to that policy.

Were these attitudes related to behavior? The policy never came to a vote; however, we could approximate an answer to that question by correlating our attitudinal measures with the person's voting intention (BI). Those correlations were quite strong, suggesting a model that would have good predictive validity.

Box 4.2 TRA/TPB Model of Participation in Waterfowl Hunting

A TRA/TPB model of waterfowl hunting participation can be developed that helps us answer the question raised in the introduction of this chapter. The model proposes that the decision to participate in hunting for a given season is influenced by three primary variables: attitude toward the behavior(s), normative influences, and perceived behavioral control (Fig. 4.2). One's attitude toward a behavior is an evaluation about performing it, e.g., one has a positive, neutral, or negative evaluation toward going hunting during the season (ATT). That attitude is a function of beliefs associated with going hunting. People might believe that if they go hunting during the season, it is highly likely that they will harvest waterfowl and that harvesting waterfowl is a highly desirable outcome. This would contribute to a positive attitude toward participation. However, people may feel they have strong constraints to their participation and little ability to control that (PBC).

A person might, for example, believe that his boss has high demands for his time, which is judged to be a very undesirable outcome and which would contribute toward a negative evaluation toward participation. Moreover, the person may have a friend or group of friends with whom he hunts regularly (SN). He may believe these friends have a strong expectation for him to participate in waterfowl hunting. The person's

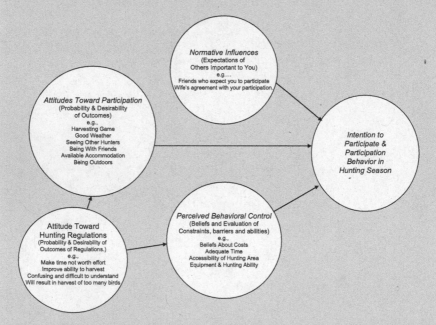

Fig. 4.2 Model of effects of regulations on waterfowl hunting participation using theory of planned behavior.

belief about his friends' expectations could have a strong influence on his decision to participate (even though the his attitude toward participation may be weak).

In the attitude formation process, the person deliberates on all outcomes that come to mind regarding the decision, weighs positives and negatives, and arrives at an overall intention. Given this model, how would regulations affect the decision to participate in the future? The simple answer is that they will only affect future participation if they become salient enough to enter the deliberative process described above.

Prior research suggests that regulations will affect hunter participation in two primary ways. First, Heberlein and Kuentzel (2002) suggest that knowledge about regulations operates by affecting a person's belief about the likelihood of harvest. Unless the attitude is already quite positive toward participation, the "new" information about regulations could weaken a person's attitude, leading to the decision to decline participation. Interestingly, the same regulation may make another person believe that fewer people will hunt and that will have a positive effect on his or her harvest potential. This would actually increase the positive attitude. The effect that a regulation will have is dependent upon the conclusions hunters draw about the effects of the regulatory change and their prior beliefs.

Second, regulatory changes can affect one's perceived ability to participate. Several studies have noted that perceptions of constraints are important determinants of participation. Enck, Swift, and Decker (1993) for example, found regulatory complexity is a barrier to waterfowl hunters in New York. Barro and Manfredo (1996) found that when deer seasons were reduced to 3 days in Colorado, hunters believed it left too little time to hunt. If the perceptions of constraints rise to a sufficient level, it tips the behavioral decision against participation.

Other relevant conclusions can be drawn from this model. First, not all people value the same outcomes; hence, there will be differences in how people are affected by increasing regulatory restriction. As noted above, some may value highly restrictive regulations because they believe it would reduce crowding, adding strength to their positive attitude. Second, those on the fence with only slightly positive attitudes toward participation, with significant constraints or without strong normative support, will certainly be the ones most influenced by increased regulatory restriction. These people are more likely to decline participation as more restrictive regulations are introduced (or increase participation as more liberal regulations are introduced).

Third, it is possible for regulatory changes to have elements that are believed by hunters to result in positive outcomes as well as negative elements. Furthermore, unless regulations are extremely prohibitive and enduring, any negative effects are likely to be reversible in the short term as the demand pool of prior hunters reconsider participation on an annual basis.

References

Ajzen, I. (1985). From intentions to actions: A theory of planned behavior. In J. Kuhl, & J. Beckman (Eds.), *Action-control: From cognition to behavior* (pp. 11–39). Heidelberg: Springer.

Ajzen, I., & Fishbein, M. (1980). *Understanding attitudes and predicting social behavior.* Englewood Cliffs, NJ: Prentice-Hall.

Albarraćin, D., Johnson, B. T., Fishbein, M., & Muellereile, P. A. (2001). Theories of reasoned action and planned behavior as models of condom use: A meta-analysis. *Psychological Bulletin, 127*(1), 142–161.

Allport, G. W. (1954). The historical background of modern social psychology. In G. Linzey (Ed.), *Handbook of social psychology* (Vol. 1, pp. 3–56). Cambridge, MA: Addison-Wesley.

Anderson, J. R. (1983). *The architecture of cognition.* Cambridge, MA: Harvard University Press.

Armitage, C. J., & Conner, M. (2001). Efficacy of the theory of planned behaviour: A meta-analytic review. *British Journal of Social Psychology, 40*, 471–499.

Bagozzi, R. P., & Warshaw, P. R. (1990). Trying to consume. *Journal of Consumer Research, 17*, 127–140.

Barro, S. C., & Manfredo, M. J. (1996). Constraints, psychological investment, and hunting participation: Development and testing of a model. *Human Dimensions of Wildlife, 1*(3), 42–61.

Bem, D. J. (1972). Self-perception theory. In L. Berkowitz (Ed.), *Advances in experimental social psychology* (Vol. 6, pp. 1–62). New York: Academic Press.

Bower, G. H. (1981). Mood and memory. *American Psychologist, 36*, 129–148.

Bright, A. D. (1997). Attitude-strength and support of recreation management strategies. *Journal of Leisure Research, 29*(4), 363–379.

Bright, A. D., & Manfredo, M. J. (1995). The quality of attitudinal information regarding natural resource issues: The role of attitude-strength, importance, and information. *Society and Natural Resources, 8*(5), 399–414.

Case, D. J. (2004). Waterfowl hunter satisfaction think tank: Understanding the relationship between waterfowl hunting regulations and hunter satisfaction/participation, with recommendations for improvement to agency management and conservation programs. Final Report to the Wildlife Management Institute for USFWS Multi-state conservation grant # DC–M-15-P, Wildlife Management Institute: Washington.

Charng, H., Piliavin, J. A., & Callero, P. L. (1988). Role identify and reasoned action in the prediction of repeated behavior. *Social Psychology Quarterly, 51*, 303–317.

Cohen, J. B., & Reed, A., II. (2006). A multiple pathway anchoring and adjustment (MPAA) model of attitude generation and recruitment. *Journal of Consumer Research, 33*, 1–14.

Conner, M., Sparks, P., Povey, R., James, R., Shepard, R., & Armitage, A. (2002). Moderator effects of attitudinal ambivalence on attitude-behavior relationships. *European Journal of Social Psychology, 32*, 705–718.

Cooke, R., & Sheeran, P. (2004). Moderation of cognition-intention and cognition-behavior relationships: A meta-analysis of properties of variables from the theory of planned behaviour. *British Journal of Social Psychology, 43*, 159–186.

Cunningham, W. A., Preacher, K. J., & Banaji, M. R. (2001). Implicit attitude measures: Consistency, stability, and convergent validity. *Psychological Science, 12*, 163–170.

Dovidio, J. F., Kawakami, K., & Gaertner, S. L. (2002). Implicit and explicit prejudice and interracial interaction. *Journal of Personality and Social Psychology, 82*, 62–68.

Duckworth, K. L., Bargh, J. A., Garcia, M., & Chaiken, S. (2002). The automatic evaluation of novel stimuli. *Psychological Science, 13*(6), 513–519.

Duda, M. D., Bissell, S. J., & Young, K. C. (1998). *Wildlife and the American mind: Public opinion on and attitudes toward fish and wildlife management.* Harrisonburg, VA: Responsive Management.

Eagly, A. H., & Chaiken, S. (1993). *The psychology of attitudes.* Fort Worth: Harcourt.

Enck, J. W., Swift, B. L., & Decker, D. J. (1993). Reasons for decline in duck hunting, insights from New York. *Wildlife Society Bulletin, 21*(1), 10–21.

Fazio, R. H. (1995). Attitudes as object-evaluation associations: Determinants, consequences, and correlates of attitude-accessibility. In R. E. Petty, & J. A. Krosnick (Eds.), *Attitude strength: Antecedents and consequences* (pp. 247–283). Mahwah: Erlbaum.

Fazio, R. H., Chen, J., McDonel, E. C., & Sherman, S. J. (1982). Attitude accessibility, attitude-behavior consistency, and the strength of the object-evaluation association. *Journal of Personality and Social Psychology, 47*, 277–286.

Festinger, L. (1957). *A theory of cognitive dissonance.* Evanston, IL: Row Peterson.

Fishbein, M., & Ajzen, I. (1975). *Belief, attitude, intention, and behavior: An introduction to theory and research.* Reading, MA: Addison-Wesley.

Gawronski, B., & Boenhausen, G. V. (2006). Associative and prepositional processes in evaluation. An integrative review of implicit an explicit attitude change. *Psychological Bulletin, 132*, 692–731.

Glasman, L. R., & Albarracín, D. (2006). Forming attitudes that predict future behavior: A meta-analysis of the attitude-behavior relation. *Psychological Bulletin, 132*(5), 778–822.

Greenwald, A. G. (2004). Revised top-ten list of things wrong with the IAT. Paper presented at Society for Personality and Social Psychology Meeting, Austin, Texas.

Greenwald, A. G., & Banaji, M. R. (1995). Implicit social cognition: Attitudes, self-esteem, and stereotypes. *Psychology Review, 102*, 4–27.

Greenwald, A. G., McGhee, D. E., & Schwartz, J. L. K. (1998). Measuring individual differences in implicit cognition. *Journal of Personality and Social Psychology, 74*, 1464–1480.

Gross, S. R., Holtz, R., & Miller, N. (1995). Attitude certainty. In R. E. Petty, & J. A. Krosnick (Eds.), *Attitude strength: Antecedents and consequences* (pp. 215–246). Location: publisher.

Haider, W. (2002). Stated preference and choice models – A versatile alternative to traditional recreation research. In A. Arnberger, & C. Brandenburg (Eds.), *Monitoring and management of visitor flows in recreational and protected areas conference proceedings* (pp. 115–121). Vienna: Bodenkultur University, Institute for Landscape Architecture and Landscape Management.

Heberlein, T. A. (1973). Social psychological assumptions of user attitude surveys: The case of the wildernism scale. *Journal of Leisure Research, 5*(3), 18–33.

Heberlein, T. A., & Kuentzel, W. F. (2002). Too many hunters or not enough deer? Human and biological determinants of hunter satisfaction and quality. *Human Dimensions of Wildlife, 7*(4), 229–250.

Heider, F. (1946). Attitudes and cognitive organization. *Journal of Psychology, 21*, 107–112.

Jackson, M. S., White, G. N., & Schmierer, C. L. (1996). Tourism experiences within an attributional framework. *Annals of Tourism Research, 23*(4), 798–810.

Judd, C. M., & Krosnick, J. A. (1989). The structural basis of consistency among political attitudes: Effects of political expertise and attitude importance. In A. R. Pratkanis, S. J. Breckler, & A. G. Greenwald (Eds.), *Attitude structure and function* (pp. 99–128). Hillsdale, NJ: Erlbaum.

Katz, D. (1960). The functional approach to the study of attitudes. *Public Opinion Quarterly, 24*, 163–204.

Krosnick, J. A., & Petty, R. E. (1995). Attitude strength: An overview. In R. E. Petty, & J. A. Krosnick (Eds.), *Attitude strength: Antecedents and consequences* (pp. 1–24). Mahwah, NJ: Lawrence Erlbaum Associates.

Lang, P. J., Bradley, M. M., & Cuthbert, B. N. (1990). Emotion, attention, and the startle reflex. *Psychological Review, 97,* 377–395.

Louviere, J. J., Hensher, D. A., & Swait, J. (2000). *State choice methods.* United Kingdom: Cambridge University Press.

Maio, G. R., & Olson, J. M. (2000). What is a value-expressive attitude? In G. R. Maio, & J. M. Olson (Eds.), *Why we evaluate: Functions of attitude* (pp. 249–270). Mahwah, NJ: Erlbaum.

Manfredo, M. J., Teel, T. L., & Bright, A. D. (2004). Application of the concepts of values and attitudes in human dimensions of natural resources research. In M. J. Manfredo, J. J. Vaske, D. Field, & P. J. Brown (Eds.), *Society and natural resources: A summary of knowledge prepared for the 10th International Symposium on Society and Natural Resources.* Jefferson, MO: Modern Litho.

Manfredo, M. J., Yuan, S., & McGuire, F. (1992). The influence of attitude accessibility on attitude-behavior relationships: Implications for recreation research. *Journal of Leisure Research, 24*(2), 157–170.

McGuire, W. J., & McGuire, C. V. (1991). The content, structure, and operation of thought systems. In R. S. Wyer, Jr., & T. K. Srull (Eds.), *Advances in social cognition* (Vol. 4, pp. 1–78). Hillsdale, NJ: Erlbaum.

Miller, D. C. (2002). *Handbook of research design and social measurement* (6th ed.). Thousand Oaks, CA: Sage Publications.

Nosek, B. (2004). *The relationship between implicit and explicit attitudes.* Paper presented at Society for Personality and Social Psychology meeting, Austin TX.

Oskamp, S., & Schultz, P. W. (2005). *Attitudes and opinions* (3rd ed., p. 578). Mahwah, NJ: Lawrence Erlbaum Associates.

Ottaway, S. A., Hayden, D. C., & Oakes, M. A. (2001). Implicit attitudes and racism: Effects of word familiarity and frequency on the implicit association test. *Social Cognition, 19,* 97–144.

Pate, J., Manfredo, M., Bright, A. D., & Tishbein, D. G. (1994). *Colorado residents' attitudes and perceptions toward reintroduction of the gray wolf into Colorado.* Ft. Collins: Colorado State University, Human Dimensions in Natural Resources Unit.

Pate, J., Manfredo, M. J., Bright, A. D., & Tischbein, G. (1996). Coloradan's attitude toward reintroducing the gray wolf into Colorado. *Wildlife Society Bulletin, 24*(3), 421–428.

Petty R., & Krosnick, J. A. (Eds.). (1995). *Attitude strength: Antecedents and consequences.* Location: Publisher.

Petty, R. E., Briñol, P., & DeMarree, K. G. (2007). The meta-cognitive model (MCM) of attitudes: Implications for attitude measurement, change and strength. *Social Cognition, 25* (5), 657–686.

Petty, R. E., & Cacioppo, J. T. (1986). The elaboration likelihood model of persuasion. In L. Berkowitz (Ed.), *Advances in experimental social psychology* (Vol. 19, pp. 123–205). Location: Publisher.

Petty, R. E., & Wegener, D. T. (1998). Matching versus mismatching attitude functions: implications for scrutiny of persuasive messages. *Personality and Social Psychology Bulletin, 24,* 227–240.

Rudman, L. A. (2004). Sources of implicit attitudes. *Current Directions in Psychological Science, 13,* 79–82.

Sanbonmatsu, D. M., & Fazio, R. H. (1990). The role of attitudes in memory-based decision making. *Journal of Personality and Social Psychology, 59,* 614–622.

Schwartz, S. H., & Tessler, R. C. (1972). A test of a model for reducing measured attitude-behavior discrepancies. *Journal of Personality and Social Psychology, 24,* 225–236.

Schwarz, N., & Bohner, G. L. (2001). The construction of attitudes. In A. Tesser, & M. Schwarz (Eds.), *Blackwell handbook of social psychology: Intraindividual processes* (Vol. 1, pp. 436–57). Oxford: Blackwell.

Sheppard, B. H., Hartwick, J., & Warshaw, P. R. (1988, December). The theory of reasoned action: A meta-analysis of past research with recommendations for modifications and future research. *The Journal of Consumer Research, 15*(3), 325–343.

Sherif, M., & Hovland, C. I. (1961). *Social judgment: Assimilation and contrast effects in communication and attitude change*. New Haven, CT: Yale University Press.

Smith, E. R., & DeCoster, J. (2000). Dual process models in social an cognitive psychology: Conceptual integration an links to underlying memory systems. *Personality and Social Psychology Review, 4*, 108–131.

Smith, M. B., Bruner, J. S., & White, R. W. (1956). *Opinions and personality*. New York: Wiley.

Stewart, W. P. (1992). Influence of the onsite experience on recreation experience preference judgments. *Journal of Leisure Research, 24*(2), 185–198.

Teel, T. L., Bright, A. D., Manfredo, M. J., & Brooks, J. J. (2006). Evidence of biased processing of natural resource-related information: A study of attitudes toward drilling for oil in the Arctic National Wildlife Refuge. *Society and Natural Resources, 19*, 1–17.

Thompson, E. P., Kruglanski, A. W., & Spiegel, S. (2000). Attitudes as knowledge structures and persuasion as a specific case of subjective knowledge acquisition. In G. R. Maio, & J. M. Olson (Eds.), *Why we evaluate: Functions of attitude*, (pp. 59–96). Location: Publisher.

Triandis, H. C. (1980). Values, attitudes, and interpersonal behavior. In H. E. Howe, Jr., & M. M. Page (Eds.), *Nebraska symposium on motivation, 1979* (Vol. 27, pp. 195–259). Lincoln: University of Nebraska Press.

Tversky, A., & Kahneman, D. (1974). Judgment under uncertainty: Heuristics and biases. *Science, 185*(4157), 1124–1131.

Wallace, D. S., Paulson, R. M., Lord, C. G., & Bond, C.F. Jr. (2005). Which behaviors do attitudes predict? Meta-analyzing the effects of social pressure and perceived difficulty. *Review of General Psychology, 9*, 214–227.

Warshaw, P. R., Sheppard, B. H., & Hartwick, J. (1982). The intention and self-prediction of goals and behaviors, In R. P. Bagozzi (Ed.), *Advances in communication and marketing research*. Greenwich, CT: JAI Press.

Wicker, A. W. (1969). Attitudes versus actions: The relationship of verbal and overt behavioral response to attitude objects. *Journal of Social Issues, 25*(4), 41–78.

Wilson, T. D., Lindsey, S., & Schooler, T. Y. (2000). A model of dual attitudes. *Psychological Review, 107*(1), 101–126.

Wood, W., Rhodes, N., & Biek, M. (1995). Working knowledge and attitude strength: An information-processing analysis. In R. E. Petty, & J. S. Krosnick (Eds.), *Attitude-strength: Antecedents and consequences* (pp. 283–314). Mahwah, NJ: Erlbaum.

Chapter 5
Norms: Social Influences on Human Thoughts About Wildlife

Contents

Introduction

During the mid-1990s I moved from the suburbs of Fort Collins, Colorado, to a subdivided farm just outside of town. The small development had a dozen, 10-acre lots on ground that was previously grazed by livestock. I soon realized that most residents in this community disagreed on what was the appropriate treatment of wildlife.

One resident built a hen house and began raising chickens. Soon, the local pack of coyotes found the hen house and began raiding it. The resident shot the coyotes from his back porch, which infuriated another resident who loved the sound of coyotes in the neighborhood.

The coyotes were attracted by cotton-tail rabbits, and the abundance of rabbits in the newly built community made it difficult to grow grass, plants,

M.J. Manfredo, *Who Cares About Wildlife?*,
DOI: 10.1007/978-0-387-77040-6_5, © Springer Science+Business Media, LLC 2008

and trees. Three of my neighbors responded differently to this. One neighbor tried to eradicate the rabbits by shooting them whenever he saw them in his yard. Another neighbor, though equally frustrated with the rabbits, could not bring himself to hurt them. The third neighbor, who loved having the rabbits in his yard, fumed about the shootings. These three neighbors never complained to law enforcement about these issues, nor would they confront one another about their differences. They believed that such confrontations would further deteriorate the sense of community among the group, and thus threaten social norms.

Social norms, or group-held rules of acceptable behavior in social life, have power. In this example, strangers were thrust together as a community when they moved next door to each other. The meaning of community varies with one's upbringing; however, there are many social rules that encourage collaboration and cooperation among neighbors. Norms of community living (things one "ought" to do) might include things like helping your neighbor with a task, waving when you pass your neighbor on the street, or socializing via community picnics. Hence, in forming a new community, this group of people would be guided by broad overarching norms about appropriate neighborly behavior. These norms explain why direct conflict about wildlife occurred rarely, even when disagreements were common.

Although the people I just described agreed on community-living norms, they did not agree on situation-specific norms: How should you keep the area around your house? How should you deal with wildlife? Should you tell your neighbors how they should maintain their areas? On these types of questions, the answers clearly differed.

Members of a given social group can enforce norms by dispensing sanctions (such as dirty looks or direct statements). If there had been wider acceptance of norms regarding treatment of wildlife (i.e., it is wrong to shoot wildlife in a community setting), then neighbors might have succeeded in shaming others into tolerating the wildlife impacts. However, if people do not agree on acceptable forms of behavior, the sanction is ineffective because it cannot provoke shame or guilt.

Why do norms differ? While researchers debate this, the answer lies partly in the fact that different people have different values. Norms are rules that are intended to ensure outcomes which groups of people value. For example, the person who believed it was acceptable to shoot wildlife from his porch valued protection of property and valued having the freedom to do what he would like on his own property. Values held by the other neighbors might involve respect for life or humaneness.

Those with similar norms formed sub-groups within the community based on the similarity of normative rules and related values. The principal division was between a group that espoused that each person should do whatever he or she wants (individualistic) and another group that believed in abiding by collectivist standards.

This chapter reviews the norm concept and discusses its importance in the area of human–wildlife relationships. Norms are important because they help us predict and understand the behavior of individuals and groups. They also set the context for conflict among groups interested in wildlife and wildlife-related activities.

Social Norms

Horne (2001a) contended that "No concept is invoked more often by social scientists in the explanation of human behavior than 'norm'" (p. 3). Norms are important because they help explain the power of the social group over the actions of individuals. They are also the foundation for interpersonal behavior among humans. As noted by Hecter and Opp (2001), "Without norms, it is hard to imagine how interaction and exchange between strangers could take place at all" (p. xi).

In the early twentieth century, norms emerged as a central concept of the classical and functional theories in anthropology and sociology (Hecter & Opp, 2001). During that period, norms were a particularly useful descriptive device in attempts to characterize cross-cultural differences of human customs and behavioral patterns. As these fields shifted direction in the mid-to-late 1900s, the use of the norm concept as a theoretical, explanatory concept received less attention. During this period of relative neglect, norms were criticized as an imprecise conceptualization (Krebs & Miller, 1985) and useful only as a post hoc explanation of behavior (Darley & Latane, 1970). This is not to imply that the terminology of norms was abandoned; in fact, it was just the opposite. Norms continued to be a regularly used descriptive term in the discourse of the social sciences. The 1990s brought a renewed interest in the concept of norms (Cialdini & Trost, 1998; Hecter & Opp, 2001). This interest was sparked, in part, by the attention given to applications of the concept in disciplines such as economics, political science, international relations, and health. The main thrust of this interest has been in attempting to incorporate the influences of social groups in explaining the behavior of individuals.

Defining Social Norms

Although specific definitions vary, social norms have generally been described as "ought" statements or rules that direct people's behavior. As noted by Coleman (1990), social norms specify what actions are regarded by a set of persons as proper or improper conduct. They not only are expectations of how one should behave, but also form expectations of how others should behave. Hence, they create predictability in the interactions of members of a social group.

Social norms are associated with specific social networks or social groups. That is, while individuals may possess beliefs about appropriate behavior, these beliefs are not norms unless they are consciously shared by a social group. From the perspective of the group, these norms tend to regulate and control the behaviors of its members and ensure their cooperation. From the perspective of the individual, one must perceive oneself to be part of a particular group and recognize that a given behavior is expected by others in that group.

The interaction of group members is an essential component of sustaining norms. As noted by Coleman (1990), a norm concerning a specific action exists when the socially defined right to control the action is not held by the actor but by others in the group. Those holding a norm claim a right to apply a sanction to others who violate the norm and recognize the right of other group members to do so as well. Sanctions would include both formal (e.g., written, legal) and informal responses toward an individual in order to enforce a given norm. Social norms are typically enforced by informal sanctions in which other members of the group may witness violation of or adherence to a specific norm and respond to the actor with disapproval or approval.

There is undoubtedly a diverse array of sanctions ranging from subtle emotional display (e.g., show of disgust or acceptance) to verbal statements to physical engagement. Sanctioning may work in several ways. First, group members may witness behavior related to a norm and directly sanction a person's behavior. For example, a hunter who sees her companion cross a fence without unloading their gun may make mention of the risk to the person. The admonishment of the friend would hopefully lead to a correction of the behavior in the future. A second way that the norm may have an effect is by the person's perception that "imaginary others" are judging the person's behavior. In this case, the threat of sanction precedes the actual act and has the effect of controlling behavior. Finally, a person may internalize a norm as a personal standard of performance and act accordingly. The process of internalization suggests that, over time, the person begins to hold the norm as a personal value or attitude (Horne, 2001b).

Multiple Social Groups, Roles, and Norms

A person certainly identifies with many social groups and these all exist under the broader umbrella of one's self-concept (Deaux, 1996). That is, they might see themselves as having an identity associated with their occupation, an identity associated with their leisure pursuits, an identity as a member of a family, etc. Each of these identities and groups is likely to evoke a specific set of normative rules that give guidance within the context of belonging to that group. Moreover, within groups, people also have distinct roles that they might assume and with that role comes another set of expectations and norms (Turner, Hogg, Oakes, Reicher, & Wetherell, 1987). For example, members of an animal rights group might all share norms about how animals should be treated. However, an elected leader of that group has an additional set of rules such as "represent the group in what I say," "provide a good role model," and "be first to sanction inappropriate behavior of members."

Symbolic interaction theorists view one's "global self" as a collection of hierarchically ranked identities with each identity corresponding to a separate social role (Rosenberg, 1979; Stryker, 1980). A role identity would be an individual's manifestation of a given social role (e.g., person X's version of being a good bow hunter).

One can readily think of groups by which people classify themselves relative to human–wildlife relationships and the hierarchical clusters that might be apparent for an individual. Groups could include hunters (muzzle loader, archer, rifle, meat, trophy, big game, small game, etc.), anglers (fly anglers, lure anglers, bait anglers, trout, saltwater, etc.), birders, managers, biologists, wildlife rehabilitators, environmental educators, scientists, and PETA members. Once a person has categorized oneself in a particular group, that person views other members favorably and sees similarity among group members and one's self. Turner et al. (1987) describe this as "self-stereotyping." A process of projection about others in the group contributes to perceptions of in-group homogeneity and in-group cooperation (Robbins & Krueger, 2005).

Through various forms of communication among members, the prescribed norms (beliefs of the ideal group member) are conveyed (Terry, Hogg, & White, 2000). Gintis (2003) suggested we can summarize the broad array of influences that lead people to internalize norms via three transmission routes: Vertical transmission is from parent figure to child; oblique transmission occurs through broader social institutions such as religion, government, schools, and media; horizontal transmission occurs via peer interactions. The actual methods of conveyance might be (a) from active instruction, stories, myths; (b) passive, via nonverbal imitation; and (c) inferred through behavior around us (Cialdini & Trost, 1998).

Once norms are learned, the more a person wants to be identified by a particular social classification, the more likely that person is to abide by the norms of that group (Deaux, 1996). A norm will affect an individual's behavior contingent upon the strength of one's motivation to identify with a group, the relevance of that group for a given situation, and the extent to which the norm is key to the identity of the group (Christensen, Rothgerber, Wood, & Matz, 2004).

Norms as Conditional and Ambiguous

While some approaches imply that norms are hard-and-fast rules, other theories emphasize their situational nature. Hecter and Opp (2001, p. 405) state that "most norms – perhaps all of them – are conditional," emphasizing the importance of understanding the scenarios in which norms do and do not operate. For example, the strength of norms governing behavior at one's area of residence might be diminished when that person is traveling away from home as a tourist.

Miller and Prentice (1996) explained conditionality by proposing that "each occasion of self-evaluation brings into existence its own frame of reference, and thus involves the construction of a standard of comparison specific to that occasion" (p. 800). Because evoked representations (thoughts and images) vary with context, so do norms. Fine (2001) suggested that norms are rule-like beliefs that are tied to values; however, they are not simply obeyed, they are "enacted." That is, people do not blindly act in accordance with a normative rule, they use it in selecting an action in their own unique situation. Fine (2001) contended that norms are conditional, ambiguous, and often take shape via negotiation or discussion among actors. For example, I was in a group of people who were in the backcountry and we were tired and cold. When stopped for a rest, we debated building a fire. The group was aware of a norm that suggests recreational fires should be avoided in the backcountry due to resource impacts and negative impacts to the experiences of other users. The discussion searched for conditional clarification of building the fire (fire to get warm is different from fire as a tradition, occasional fire versus routine as part of a camp, small fires versus large fires, etc.). The group negotiated a position that suggested that, on this occasion, a small fire was acceptable.

Fine (2001) contended that an important component of enacting a norm is the way in which it is framed. A frame is a template for understanding particular events, circumstances, and the actions of others. Frames are a representation of meaning matched with previous experiences and the contexts in which they occurred. The behavior actually enacted is the result of one's search for the appropriate frame and an application of this frame to the specific demands of the situation. Fine (2001) illustrated these notions in a study of an amateur organization of mushroom gatherers. According to Fine (2001), mushroomers are careful to frame their activities so that they are consistent with their belief that the woods should be protected from human intrusion. Hence, they frame their activities as minimizing harm and they differentiate themselves from those that do more damage. Norms draw distinctions between proper and improper behaviors within this frame.

To summarize, norms are prevalent in directing our day-to-day activities. They are rules about acceptable behavior that have real or imagined sanctions. Norms are associated with being a member of a certain group, a role within that group, and the extent to which one associates his or her identity with either the group or the role. While at times norms might be hard-and-fast rules, they are often used by people as guidelines for interpreting a situation and choosing among various behavioral alternatives. In the next section, I discuss some of the reasons for the prevalence of norms in daily life.

The Origins and Emergence of Norms

Why are norms such a powerful force in social situations? One can deal with this question at two levels. At the most basic level we might ask: Why have norms become part of all human cultures? At another level we might ask, how does one specific norm emerge and become enacted over other possible norms?

Aberle, Cohen, Davis, Levy, and Sutton (1950) suggested that norms are not simply important, they are a prerequisite to a functional society. More specifically, they contended that society requires a shared, articulated set of goals and normative regulation of the means to these goals. The corollary of this has been expressed by Eriksen (1995) who stated that norms are a reflection of the basic values of society and that the types of sanctions applied to different types of norm-breaking give an idea of the relative importance and power associated with the different values.

Theorists with an evolutionary perspective see the emergence of norms as critical in human adaptation to environmental surroundings. Kenrick, Ackerman, and Ledlow (2003) stated, "what humans are inclined to learn, what humans are inclined to think about, and the cultural norms that humans create are all indirect products of the adaptive pressures that shaped the human mind" (p. 111). To further illustrate, Fehr and Fischbacher (2004) observed that "human societies represent a spectacular outlier with respect to all other animal species because they are based on large-scale cooperation among genetically unrelated individuals" (p. 185). They contended such cooperation is possible due to the ability of humans to establish and enforce norms and that the human characteristics that led to social cooperation emerged due to the selective processes of evolution. Norms leading to cooperative behaviors were selected due to advantages provided in mating, acquiring food, defending against threats, etc.

Perhaps a more practical question asks why one type of norm has emerged over another alternative. Much of the current debate about the emergence of norms revolves around whether they are "backward looking" or "forward looking" (Elster, 1989). The backward-looking view emerges from work by Emile Durkheim, Margret Meade, and others who emphasize that norms are held now because they were in the past, are part of the tradition of a given social group, and help explain group solidarity (Cialdini & Trost, 1998; Elster, 1989). This view is consistent with the explanation that norms arise due to regularity of behavior (Horne, 2001b). A person may initially engage in a behavior because it provides some advantage. As the behavior occurs more regularly, it is imitated by others. As the behavior is repeated over and over, there is an "oughtness" associated with it as it becomes expected. When people act outside the expectation, they are sanctioned. The more common the behavior, the more it is expected and deviations are punished.

For example, farmers in the western United States commonly burn dead vegetation that grows up around hedgerows, ditches, or areas along roadsides. Perhaps the original purpose for such action was to burn back weeds; however, over time, the practice has become commonplace. With the re-growth of bright green grass following the fire, the burned areas would be viewed as attractive and an indication of a well-kept farm area. Through such a process, burning ditches became a norm among the farmers in the western region – a norm that had unfortunate impacts on many species of wildlife. For many species, ditch areas were important places for nesting, resting, and thermal protection. Without heavy vegetation, these areas become unsuitable for such purposes.

The forward-looking view of norms suggests they emerge because they hold the promise of some future reward (or avoidance of negatives) for a group. Hechter (1987) proposed that groups exist primarily for the point of providing their members some joint good. These goods can only be attained if members comply with the rules, i.e., norms that ensure the delivery of these goods. That is, norms arise because they provide rewards or advantages (minimize disadvantages) for group members through the cooperative behavior they require, even though specific individuals may have to exhibit self-sacrifice. Hence, norms exist in our day-to-day world in order to achieve the goals of a given group and ensure for that group the values that it seeks. This *instrumentality proposition* states that "if members of a group have a goal and they believe that a norm is instrumental for attainment of that goal, it is likely that the norm emerges" (Coleman, 1990, p. 242).

To illustrate, fly anglers value the opportunity for solitude and the challenge to cast dry flies to rising fish. Norms among these anglers tend to preserve this valued situation. Norms have arisen that suggest it is inappropriate to enter the stream too close to a person already fishing at a spot, even though the person might be catching fish. Norms would also dictate that if a person is casting to rising fish near the bank of the stream, people should make all efforts to avoid walking close to the bank for fear of scaring the fish. Those adhering to the norm (sacrificing in those instances) have the expectation that when they are in a similar situation, norms will preserve their attainment of a quality fishing experience.

Critics have levied several challenges for the instrumentalist approach. For example, it is pointed out that many existing norms do not in fact benefit a group and in some cases are detrimental to it. As noted by Elster (2003), "Some norms do not benefit anyone, but are rather sources of pointless suffering" (p. 297). Elster (1989) also criticized the instrumental view for holding untenable assumptions that include (a) people have advanced knowledge of the effects of a norm, (b) they know how they can contribute to a norm's enactment, (c) all individuals have equal interest (benefits) in a norm, and (d) there are effective ways to deal with "free-riders" (i.e., those that disregard the norm and act in self-interest). Opp (2001) noted that most contemporary views of norms incorporate an instrumental component; however, he also suggested that the assumptions for a pure functionalist explanation of norms, as identified above by Elster (1989), are very difficult to meet. He suggested that instrumentality is but one of several factors that affect the emergence of norms. Norms will arise based on imperfect, subjective knowledge, which may give rise to norms that are advantageous to a minority of individuals or for just a short period of time. While it may be difficult to provide an instrumental explanation for every norm imaginable, it is hard to deny the instrumentality of norms in a great many cases of social behavior.

Conceptual Approaches in Application of Norms

A number of different theorists have used the concept of norms in an attempt to understand and predict human behavior. In this section, a few of the prominent approaches are briefly summarized. Consistent with the primary thrust of the approach advanced in this book, the theories reviewed here are focused on the influence of norms on the behavior of individuals.

Schwartz's Normative-Based Decision Model

Schwartz applied the concept of norms in a theory of moral decisions (Schwartz, 1968; Schwartz, 1977; Schwartz & Howard, 1982). This includes decisions where actions are taken that affect the welfare of others, the person taking the action is perceived as responsible, and the behavior is deliberate. Schwartz (1968, 1977) proposed that personal norms are highly influential in the moral decision process. According to Schwartz and Howard (1982), personal norms are "situation-specific behavioral expectations generated from one's own internalized values, backed by self administered sanctions and rewards" (p. 329). The theory suggests that it is these self-imposed views of what is right or wrong that will dictate one's behavior. Schwartz and Howard (1982) distinguished personal norms from social norms, which are defined as "group expectations backed by externally defined and imposed rewards and punishment" (p. 329). However, Schwartz contended that personal norms are far more influential and enduring than social norms in the moral choice context.

In Schwartz's view, an action situation is initiated when people's attention is drawn to an event that might stimulate action. In this initial phase, people evaluate the need to react, try to determine whether action is possible, and make an assessment of whether they have the ability to take action. For example, once when I was in a very remote area hunting, I came across a hunter who was sprawled out by a game trail, apparently passed out. I was unsure whether the person was injured and wondered what I could do if he were injured, as I had no advanced first aid training. It turned out that the person had been hunting and had become quite tired. He sat down to hunt and had fallen into a deep sleep from which I awoke him!

In cases where there is a perceived need and ability, a person's internal values are evoked along with specific personal norms. Schwartz (Schwartz, 1977; Schwartz & Howard, 1982) contended that personal norms will be highly influential if people are aware of the consequences of their actions and believe that they have a responsibility in the situation. For example, if the hunter I found could not be awoken, values regarding universalism would have been aroused (see Chapter 6). Given the remote situation, I would have readily assessed that the person might perish if he was not helped (consequences). Because no one else was likely to come along to help the person, it was my responsibility to help him.

Norm activation is followed by a person evaluating the costs and benefits (both normative and non-normative) of engaging in a specific behavior. For example, I might have thought to carry the injured person to safety but would perhaps consider my own safety in taking such an action. In highly conflicted situations, e.g., where costs and benefits are similar, a person may engage in a defensive action that "redefines" the situation, making one's action or inaction more clear.

Schwartz's model emphasizes the notion that personal norms are constructed and unique to each situation. Schwartz and Howard (1982) stated that personal norms are "situation-specific reflections of the cognitive and affective implications of a person's values for specific actions" (p. 337). For example, a person might see someone feeding wildlife at a national park. Assuming that the observer places high value on the natural conditions within parks, she may feel that it is wrong for humans to feed wildlife there. She may feel that something should be said to the person who is feeding the wildlife, yet she may notice that an enforcement officer is in the area and is taking no action. She must construct a norm that fits the situation. Perhaps she should not act in a situation where enforcement officers are present, or she may construct a norm that the officer should set an example of enforcement by saying something to the person feeding the wildlife. Schwartz and Howard (1982) contended that the personal norm that is constructed will depend on whether the affected value is central to one's own self-evaluation. That is, if she identifies herself strongly as a park purist, it is more likely that she will act in this situation.

Although originally introduced to explain altruistic behavior, Schwartz's theory has been applied in several environmental behavior studies. Heberlein (1972) argued that norms can be cast as a moral decision; that because the environment effects people, actions toward the environment impact people. Because of their effects on people, norms are moral decisions. Hopper and Neilsen (1991) conducted an experiment to test effects of various interventions on recycling behavior and found some support for the Schwartz model. Similarly, Bratt (1999) conducted a study of recycling behavior and determined that personal norms mediated the effect of social-norm awareness of consequences. Personal norms were strong predictors of self-reported behavior.

Theory of Reasoned Action and Theory of Planned Behavior (TRA/TPB)

Two broad sources of influence on human behavior have been discussed in the social psychological literature. One emphasizes the role of individual deliberation, such as in weighing the perceived positives and negatives of a particular behavior. The second emphasizes the influence of one's social surroundings and the pressure to conform to group norms. These two influences were brought together in Fishbein and Ajzen's Theory of Reasoned Action (TRA) (Fishbein &

Ajzen, 1975) and in a later version of TRA known as Theory of Planned Behavior (TPB) (Ajzen, 1991). This widely used theory proposes that a person's behavior is a function of attitude and subjective norm. Subjective norm was conceptualized as a person's belief about what important others want one to do and one's motivation to comply with those others. While TRA and TPB have been shown to have excellent predictive validity, the strength of prediction has been found with attitude, not subjective norm (Ajzen, 1991; Armitage & Conner, 2001; Farley, Lehmann, & Ryan, 1981). To illustrate, in a study of hunting behavior, Hrubes, Ajzen, and Daigle (2001) found that attitude was a much stronger predictor of intention to participate in hunting ($b = 0.55$), than was subjective norm ($b = 0.36$). Advocates of TRA/TPB claim the influence of subjective norm will vary by the type of behavior under consideration (Trafimow & Fisbein, 1994), the type of person involved (Trafimow & Finlay, 1996), and the extent to which one's *private self* or *collective self* has been made salient (Ybarra & Trafimow, 1998). To illustrate an interaction between type of behavior and person, Young and Kent (1985) conducted a study on how people made the decision to go camping. They found that subjective norm was a more powerful predictor for women than it was for men, suggesting that, regarding camping behavior, women may be more influenced by subjective norms than men.

Despite its predictive strength, TRA/TPB has been criticized in its use of norms on both measurement and conceptual grounds. Methodologically, TRA/TPB treats attitude and subjective norm as independent effects. That is, typically, in tests of TRA, behavioral intention is regressed on attitude and subjective norms and the beta weights are taken to reveal the relative influence of each effect. Studies, however, have indicated that theses components (attitudes and subjective norms) are likely to be correlated (Shepherd & O'Keefe, 1984). Using structural equation modeling, Vallerand, Pelletier, Desaies, Cuerrier, and Mongeau (1992) further provided evidence that the direction of impact is more likely from norms to attitudes than the reverse. In other words, while TRA/TPB proposes that norms and attitudes act separate from one another, this evidence suggests a path where norms affect attitudes which in turn affect behaviors.

Terry et al. (2000) identified further methodological problems in TRA/TPB's use of modal salient referents. In the implementation of TRA/TPB, Ajzen and Fishbein (1980) prescribed that the referents cited most frequently among a pretest group be included in the final survey instrumentation (i.e., the modal salient referents). This assumes that all people in a sample have the same set of referents and association with particular groups. Terry et al. (2000) claimed that the individual variation in group membership is important to retain when measuring referent groups.

Finally, Armitage and Conner (2001) provided a review of prior studies that shows the lower effects of subjective norms (SN) when compared to attitudes may be due to the actual SN measure employed. To clarify, TRA/TPB employs a global measure of SN (asking what important others want you to do and your motivation to comply) and an indexed SN measure formed from measures on modal salient referents. These separate measures reflect the conceptual approach shown by the following TRA/TPB formulation which suggests over-all influence of important others is a function of the influence of the individual referents that are salient:

$$SN = \sum_{i=1}^{n} b_i mc_i$$

where i is a salient referent group, b_i is one's belief that the referent would or would not want the person to perform the target behavior, mc_i is the motivation to comply with the referent, and n represents the number of modal referent groups.

Armitage and Conner (2001) showed that studies using the global SN measure have noticeably lower beta weights than the indexed measure. Because most studies use the global SN measure, the SN component might be consis-tently reported as a lower effect.

Terry et al. (2000) cited weaknesses in the social norm component of TRA/TPB due to the fact that it reflects only the injunctive component of social influence (i.e., the part that responds to the need for social approval and acceptance and the part that creates pressure to conform to others). This is different than the "informational influence" in which a person accepts and internalizes the information from others. Moreover, they contended that TRA/TPB views subjective influence as an additive function across a number of referents that an individual defines as important to them. Terry et al. (2000) contended it is more likely that norms are influenced by a single reference group that is most salient to the situation. Finally, TRA/TPB focuses on the people that are important referents, not the group of which a person is a part. Terry and colleagues (Terry & Hogg, 1996; Terry, Hogg, & White, 1999; Terry et al., 2000) have led calls for a modification of TRA/TPB that takes a broader view of the social influence process. This approach is discussed later in the chapter.

Norm Focus Theory

Cialdini, Reno, and Kallgren (1990) introduced a norm focus approach to help clarify the operation of norms. Additionally, this work highlights the impor-tance of the salience of a norm as a prerequisite to action, assuming that unless a norm is brought to the top of a person's mind, it will not have an effect (Cialdini et al., 1990). Beyond the importance of salience, the norm focus model empha-sizes the importance of distinguishing descriptive from injunctive norms. Injunctive norms reflect people's perceptions of what others want them to do. Descriptive norms are observable regularities of behavior that would provide cues regarding socially acceptable forms of behavior in a given situation.

Descriptive norms are important when we have a degree of uncertainty about appropriate action in our social life. When people are placed in uncertain situations, they seek information that helps them preform correctly. Cialdini et al. (1990) suggested that this tendency is related to a basic motivation to be effective and competent in our action. When uncertainty exists, we look for guidance in what others are doing (i.e., the descriptive norm). The concept of descriptive norm can be illustrated with an example of visitors to national parks, where norms of appropriate behavior may be vague. Visitors may be tempted to venture close to the abundant wildlife frequently found in parks, but be uncertain about the acceptability of such action. In this case, the observation of others' behavior (i.e., the descriptive norm) might be a strong influence on action. For example, if the person observes a high proportion of other visitors proceeding toward wildlife, the person would conclude that is an expected form of behavior at the location (i.e., the descriptive norm). However, if among a large group of visitors, only one person approaches wildlife, it may draw focus to the fact that few others are doing so, revealing it is not an appropriate behavior.

Injunctive norms are those with known societal approval or disapproval with a clear possibility of sanctions. Such norms are shared by the group to which a person belongs. Cialdini et al. (1990) suggested that the power of injunctive norms is related to one's motivation for social acceptance. Their research in the area of littering suggests that injunctive norms are likely to override the effect of descriptive norms in a given situation. For example, a norm against littering (particularly in national parks) is widely held in the United States. Hence, despite what cues a person might get from the descriptive norm (e.g., seeing evidence that others littered), it would be expected that the antilittering injunctive norm would have a prevailing effect if that norm was salient to the person.

A number of studies have tested the predictive validity of the descriptive norm concept within the TRA/TPB framework. Descriptive norms have been operationalized via survey items such as "Most people I know do behavior X." Using this type of approach, Conner and McMillan (1999) found that a descriptive norm variable added to explained variance in predicting cannabis use. Other studies have revealed similar findings (DeVries, Backbeir, Kok, & Dijkstra, 1995; Fekadu & Kraft, 2002; Grube, Morgan, & McGree, 1986).

Identity Theory

Identity theory has emerged from a sociological examination of social group processes and a symbolic interactionist theoretical perspective (Astrom & Rise, 2001; Charng, Piliavin, & Callero, 1988). According to this theory, one's perception of oneself is divided into multiple, hierarchical role identities. Each role identity (also referred to as self-identity) is defined by a position in a social community. It has a set of characteristics and expectations for appropriate behavior (Astrom & Rise, 2001; McCall & Simmons, 1978). These roles vary in their importance to a person and hence vary in their influence on the individual's behavior. The degree to which a person internalizes the role identity, and the degree to which it is salient to a person, defines the "role–person merger" for that person (Turner, 1978). Factors that influence the salience of a given role identity include the extent to which significant others identify the actor in the role, the amount of social support one receives for an identity, and the size of the individual's social network (as related to the role). When the group is salient, a person's participation in role-congruent behavior validates his or her self-concept and status as a group member (Astrom & Rise, 2001).

Several studies have tested the predictive ability of role-identity variables in the context of an attitude–behavior model such as TRA/TPB. For example, Astrom and Rise (2001) examined healthy eating behavior among young adults and determined that role identity explained variance beyond attitudinal, normative, and past behavior measures. Role identity was operationalized with items such as "I look at myself as a person that eats healthy food." Studies examining a variety of other behaviors have found similar supportive results (Abrams, Ando, & Hinkle, 1998; Biddle, Bank, & Slaving, 1987; Charng et al., 1988; Conner & McMillan, 1999; Granberg & Holmberg, 1990; Theodorakis, 1994).

Social Identity Theory (SIT) and Self-Categorization Theory (SCT)

Social identity has been described as that component of one's self-concept that is derived from one's knowledge of group membership and the value and emotion attached to that membership (Tajfel, 1981). Turner et al. (1987) suggested that people define and evaluate themselves in terms of distinct social categories (e.g., father, team fan, antihunter, angler). As part of the identification process, there is a tendency to perceive distinct differences between the group one belongs to and other groups. Further, there is a tendency to perceive in-group thoughts and behavior as positive, leading to one's own self-enhancement as a member. Turner et al. (1987) proposed that when identity is salient, it will influence the behaviors of individuals. The process by which this effect occurs has been explained through self-categorization.

Self-categorization theory proposes that people classify themselves as part of distinct social groups (Turner et al., 1987). When a specific situation prompts one's social membership, the person creates a group-specific norm that represents the "prototype" of a group member. The creation emanates from shared consensual information of group members. The prototype prescribes the attitudes, beliefs, feelings, and behavior for the situation. In this process, the self is transformed and represented as the group. Within this theoretical context, two key factors influence the use of social norms as behavioral guides (Christensen et al., 2004). One is the strength of a person's motivation to identify with a group while the other is situational factors that increase a person's self-categorization. To illustrate the later point, a person in the work environment might find himself in a group that is discussing whether hunting is an ethical endeavor. As the discussion proceeds (a situational factor), his identity as an antihunter may become quite salient and evoke normative views of this topic.

Terry et al. (2000) reviewed several studies that provide support for predictions borne from the SIT/SCT. The findings suggest that attitude–behavior consistency is influenced by the in-group norm present in a situation. When one's attitude is consistent with the norm, attitude–behavior consistency is high. However, when they conflict, prediction is lower. The theory also proposes that the more one identifies with a specific group, the stronger the predictive ability of the norm.

Other studies conducted within the SIT/SCT framework suggest it offers a promising approach to assessment of normative influence. Christensen et al. (2004) showed that the more a person identifies with a group, the more positive were emotions for members that conformed to the group's norms (compared to those who violated the norm). In a series of studies regarding homeless people's use of social support services, variables such as identification as a support service user and group norms provide explanation over and above other TPB variables (Christian & Armitage, 2002; Christian & Abrams, 2003; Christian, Armitage, & Abrams, 2003). Schofield, Pattison, Hill, and Borland (2001) found that smoking behavior was associated with membership in a group with a favorable group norm toward smoking. The relationship was even stronger for those who strongly identified with the group. Finally, in an interesting cross-cultural study, Jetten, Postmes, and McAuliffe (2002) suggested that those who identified strongly with their national group were more likely to endorse the norms associated with their national identity. More specifically, those who identified strongly with a collectivist culture (i.e., Indonesia) agreed with norms that emphasized the group, while individuals who identified strongly with an individualist culture (i.e., United States) agreed with normative statements that assert individuality. For example, those who identified strongly as being American were more likely to score highly on normative measures taken to represent individualism.

Norms in Natural Resource Management

The concept of norms has played a prominent role in research dealing with the human dimensions of natural resources. Most of this research has been conducted within one specific paradigm, referred to as a *structural characteristics model*, led by a group of researchers including Tom Heberlein, Bo Shelby, and

Jerry Vaske (Heberlein, 1977; Shelby, 1981; Shelby & Heberlein, 1986; Shelby & Vaske, 1991; Shelby, Vaske, & Donnelly, 1996; Vaske, Shelby, Graef, & Heberlein, 1986). The approach has had the specific purpose of aiding natural resource managers in selecting management actions and in setting impact standards (Vaske & Whittaker, 2004). The structural model is primarily descriptive and is intended to reveal modality and consensus among groups. It is based on a graphic device introduced by Jackson (1965) known as the *return potential model*.

Implementation proceeds with collection of data on a defined group of individuals relevant to a management issue (e.g., visitors to a specific area, a group of hunters, the public of a state). Questions are asked of study participants about the acceptability of an action, an encounter type or level, or a resource impact, depending on the nature of the study. Data are analyzed and displayed via the return potential model format, as is displayed in Fig. 5.1. It is critical to realize that this research tradition defines survey responses regarding acceptability as personal norms and the aggregation of all personal norms of the group being studied as social norms.

Figure 5.1 illustrates how the structural approach guides interpretation of findings. The x-axis shows the independent variable that prompts the personal norms. This figure uses data from a survey of the Denver-metro residents of Colorado regarding their normative beliefs about acceptability of management actions for dealing with conflicts with mountain lions (Zinn, Manfredo, Vaske, & Wittmann, 1998). The x-axis variable might be interval, such as increasing numbers of encounters with other recreationists, or ordinal, such as shown here in increasing severity of incident. Curves are constructed by plotting means of the acceptance against levels of the independent variable. Multiple curves can be projected onto a single graph allowing for easy comparison among a variety of things such as users or types of management actions, as is shown here. The range of acceptability includes all points above the neutral line. In this example, destroying an animal is acceptable only when the animal injured or killed a human. The intensity of a norm is defined by the distance away from the neutral line. Hence, the intensity of the social norm regarding destroying a mountain lion is much greater when it has killed a human compared to when the human was just injured. In summary, Fig. 5.1 shows that relocation of a mountain lion is acceptable to the public in all cases except death to a human; monitoring is acceptable only in cases where there is no injury or death to humans; and destroying the animal is acceptable only with death or injury to humans.

The structural approach has been used in a wide variety of natural resource contexts (see Vaske & Whittaker, 2004, for an overview) and has proven to be a highly useful tool for managers. However, the structural approach has not been without criticism. Perhaps most pertinent for the current chapter is the observation that the structural model operationalizes a significantly different notion of norm than is described earlier in this chapter. It is, in many ways, more similar to the concept of attitude described in Chapter 4. As noted by Heywood (1996,

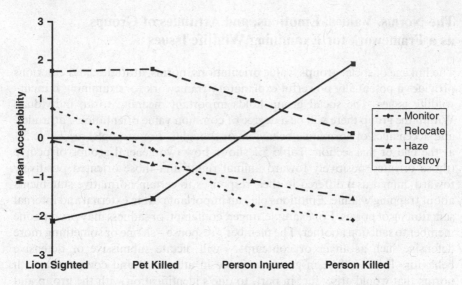

Fig. 5.1 Use of the Structural Norm Approach for Identifying the Context of Acceptability for Different Management Actions for Residents of the Denver-Metro Area, Colorado. From Zinn et al. 1998

2000), norms measured via the structural approach may not meet the criteria of the more traditional definition of norms. Manning, quoted in Heywood (2000), suggested that the use of the structural approach in recreation research has focused "on conditions rather than behaviors, they do not necessarily involve a sense of obligation on the part of the respondent, and there may be no form of sanctions to reward or punish associated behavior" (p. 262). Other concerns have also been raised about the generalizability of structural norm studies suggesting that there is little stability of findings across study locations, study populations, and methods of questioning to determine personal norms (Hall & Roggenbuck, 2002; Roggenbuck, Williams, Bange, & Dean, 1991; Williams, Roggenbuck, Patterson, & Watson, 1992). This research suggests that in many cases, study participants do not have accessible norms that pertain to the situation in question. In response to these concerns, Donnelly, Vaske, Whittaker, and Shelby (2000) introduced the notion of norm prevalence to their model. High prevalence occurs when norms are salient to the group being investigated while low prevalence implies a lack of saliency. The higher the norm prevalence, the more likely a norm exists and the more likely a manager could expect valid results when surveying the public about the norm.

While these concerns merit caution in future studies, the use of the structural approach persists due to its practical utility for management. The approach has been particularly useful in selecting social standards during the planning and decision-making process.

The Norms, Values, Emotions, and Attitudes of Groups as a Framework for Examining Wildlife Issues

The linkage of social groups, value orientations, norms, attitudes, and emotions provides a potentially powerful explanatory framework for examining human–wildlife issues. The social group holds important meaning to an individual. Within the group there will be a degree of common value orientations, attitudes, and acceptance of normative behaviors or thoughts. For example, see Table 5.1 at the end of this section. Table 5.1 shows how two general groups of people (those oriented positively toward animal rights and those oriented positively toward animal use) differed in their responses to certain normative statements about trapping wildlife. Emotions play an important part in external and internal sanctioning of norms. For example, anger, contempt, or sadness may prompt one member to sanction another. The member's response – shame or something more defensive such as anger or contempt – will dictate submissive or defensive behavior. There will be in-group variance in adherence and compliance with norms that would arise due, in part, to one's identification with the group and one's strength of value orientations or conflicts within the individual (e.g., conflict with other beliefs, conflicts with other groups). To the extent a person

Table 5.1 Differences on normative statements about wildlife by wildlife value orientations for Colorado residents

Normative statements about trapping wildlife	People oriented positively toward wildlife rights		People oriented positively toward wildlife use	
	% Agree	% Disagree	% Agree	% Disagree
Trapping wildlife is never acceptable for any reason	31.1	43.2	2.0	39.4
Trapping wildlife is acceptable to prevent the spread of disease such as rabies	48.4	8.8	41.0	2.2
It is acceptable for people to trap wildlife if it is done to prevent economic loss	1.2	38.7	40.8	7.2
It is acceptable for people to trap wildlife to protect livestock and property	45.4	22.5	37.7	2.5
It is acceptable for people to trap wildlife if it is done primarily to obtain money	11.5	75.0	15.3	18.9
It is acceptable for people to trap wildlife for recreation	3.6	70.5	5.9	32.3

This table includes only people who agreed/disagreed moderately or strongly with the statement (i.e., the *neutral* category and the *slightly* category were omitted).
Source: Fulton, D. C., Pate, J., & Manfredo, M. J. (1995). *Colorado resident's attitudes toward trapping in Colorado*. (Project Report No. 23). Project report for the Colorado Division of Wildlife. Fort Collins: Colorado State University, Human Dimensions in Natural Resources Unit.

identifies with a group and its values, the more likely they will adopt the proto-typical attitude of the group and the more likely that person will respond to and adopt group norms. Norms reinforce the values and value orientations of an individual and the group that individual belongs to. Norms would ensure behavior that is directed toward attainment of values (e.g., *never harm animals* might be the norm of a PETA member that shares strongly mutualistic values with other PETA members). Accordingly, it would be expected that differences in normative behavior regarding wildlife and its treatment vary directly by the predominant value orientations held by an individual and a group.

This provides an interesting framework for examining wildlife management and policy issues that are often framed in us-versus-them group perspectives (Brewer, 1999). The framework described here might emphasize measurement of participants' perception of and attachment to specific groups, the perceived clarity and extent of the group norms, the disparity or congruity of group norms and individual attitudes, and the centrality of values that bind its members. These group characteristics would be proposed to affect the power and effectiveness of a group in policy debates.

This type of approach would certainly add an interesting perspective to the preponderance of human dimensions of wildlife research that focuses on the individual. In most of this research, a social group is important only as a means of classifying a population of interest (e.g., hunters versus anglers, membership in organized groups such as PETA or NRA). Little research directly examines how becoming a member of a group effects the thoughts and actions of individuals. That trend merely reflects what has occurred in social psychology. Forgas and Williams (2001) stated, "During the past few decades social psychology has increasingly adopted an individualistic social cognitive paradigm that has mainly focused on the study of individual thoughts and motivations" (p. 5). Yet the increasing trend toward examination of social and group influences seems particularly relevant for issues regarding wildlife (see Box 5.1).

Box 5.1. Hunters as a Group: Illustrating how actions of the group can be explored through the integration of norms, values, attitudes, and emotions Hunting in North America might provide an excellent example of a powerful group phenomenon in human-wildlife interactions. Being a "hunter" in North American society can be very central to one's perception of self. Typically, hunters are socialized to the sport at an early age by a parent or close relative. The immediate group of hunting companions is often a critical part of one's social network. Continued participation requires a strong commitment both psychologically and economically (Barro and Manfredo 1996). Further, due in part to threats to the group's continuation from anti-hunting initiatives, there is strong in-group identification and views are strongly differentiated from non-hunters (e.g., non-hunters are viewed as misinformed or unknowledgeable about issues regarding game). Hunting is also often identified with maturation of youth and there are strong symbolic and ritualistic meanings associated with attaining group membership. Many hunters must "earn" the right (from parents and mentors) to hunt as a youth, proving their independence, responsibility and maturation. Hence, for those that identify with being a hunter, this self-classification can be very central.

There are strong and clearly defined norms that preserve the goals associated with the hunting experience. This might include goals such as harvesting and seeing wildlife (norms about waste and wounding, not exceeding limits, being quiet in the woods, not poaching or harvesting out of season), ensuring safety (norms about gun transport, gun handling, being prepared for emergencies), or maintaining the freedom and North American tradition of hunting (speaking against threats to continuation of hunting, favoring initiatives that preserve the right and opposing those that threaten it). Members are also expected to hold certain attitudes and beliefs, such as, "hunting is necessary to keep populations from growing out of control." Moreover, in efforts to promote group goals, its members actively espouse its normative beliefs to others.

Given the centrality of hunting to many of its members, it is reasonable to hypothesize that emotional response and severity of sanctions to norm violations will vary with strength of identification as a hunter. The more important one's identity as a hunter, the stronger the emotional response. Also, as we examine the effects of group membership as hunters, it is reasonable to propose that hunters are far more likely to be driven by group norms (than individual attitudes) when compared to non-hunters on topics related to wildlife.

Empirical support for this assessment would provide important insight to working with hunters groups. For example, it would suggest that the norms and attitudes held by the group are highly influential on thought and behavior its members. Strategies such as targeting normative beliefs in communication, using referent groups or opinion leaders of referent groups in communication with hunters, and the use of group leaders in stakeholder processes should be effective. Moreover, given the deep processes of recruitment to the group, it is unlikely that people would join in a casual fashion. There is a strong component of socialization that could not be replicated easily (e.g. such as programs to recruit hunters). Finally, given the strong identity and centrality of group membership, it is likely that the group could easily change but could be quickly mobilized to action in policy disputes.

In particular, this might open a line of inquiry that has implications for managers who increasingly use stakeholder processes to facilitate decision making. In implementing these stakeholder processes, we are rarely aware of the characteristics of the groups who participate. What is the size of the group? To what extent do group members identify with the group? What are the normative positions of the group and what is the communication flow among the group? How influential is the group in affecting the attitudes and intentions of its members? These topics beg further exploration in the context of norm theory.

Conclusion

Norms influence most day-to-day areas of life. They are nuts-and-bolts of social group cohesion and are driven by a strong human need for inclusion. They are not only influential in group settings but also when one is alone, imagining what others might think. Norms effect which attitudes we adopt as our own because we desire to be part of a group that holds certain views.

The conceptual structure of norms provides a useful way to explore many areas of human–wildlife relationships. Past norms research helps us understand the boundaries of acceptability on key management issues (e.g., management's treatment of wildlife, tolerances for different types of wildlife). It helps us contrast the beliefs of different stakeholder groups to understand conflict. It has also been an important component of models used to predict the behavior of individuals. Our ability to influence human behavior is enhanced by a better understanding of the operation of norms. To understand norms, we need to improve our understanding of

- How group norms influence the development of an individual's attitudes.
- How norms form within groups.
- How norms are interwoven with concepts like emotion, attitudes, and values.
- How norms emerge in concert with larger-scale phenomenon – such as demographic, technological, or institutional change.

Management Implications

Norms have offered several practical tools for fisheries and wildlife managers. A few key examples of these uses are described below.

Setting Management and Planning Standards   One of the most critical and challenging tasks for resource managers is setting standards. In the managerial realm, standards offer concrete, measurable variables that objectify the more general statements of management intent. For example, how do you quantify ideals such as a *quality recreation experience*, a *positive work environment*, or *sustainable tourism*? In all these cases, the challenge is in the details of identifying the best variables that would give an accurate indication of these ideals and then a level on that variable that clearly defines the ideal. For example, the presence of a natural environment can be indicated, in part, by the level of non-natural sounds measured in decibels (Monroe, Newman, Pilcher, Manning, Stack, 2005). The level (in decibels) that defines this would be the standard to maintain. A significant stream of social science research in natural resources has applied the concept of norms in a standard setting. This approach assumes that, in many cases, ideals (like *quality* or *sustainability*) can be defined through the social consensus identified by norms. For example, if a preponderance of people say a quality wilderness fishing experience entails seeing no more than five other groups of people on the trail in a day, then that forms the starting point for standard setting. Other considerations in the standard-setting process might include current levels of use, the practicality of enforcing standards, degree of consensus for a particular standard.

Setting Social Carrying Capacities An early application of the social sciences in natural resources was to describe recreation carrying capacity (Wagar, 1964). In the 1960s, outdoor recreation on public lands expanded dramatically. This compromised resources, facilities, and the quality of recreation; concern grew among stakeholders and managers. Several writers introduced the notion that there was a carrying capacity; they thought this carrying capacity could be measured, and that this measure could justify limiting recreation use.

Initially, researchers wanted to identify a single number, based on the number of recreationalists a resource could support, and use this number as a measure of carrying capacity. This approach was shown to be futile and in its place, researchers proposed management processes that could set capacities. These management processes illustrate the use of this new approach: The Visitor Experience and Resource Protection (VERP), Visitor Impact Management (VIM) and Limits of Acceptable Change (LAC) planning systems (see Manning, 2007, for a description of these approaches).

The prevailing approach to setting visitor capacities focuses on clarifying objectives about the types of conditions to maintain and establishing standards by using the norm approach (Manning, 2007).

Understanding the Basis for Depreciative Behavior Our behavior in uncertain situations is often directed by our inference about the norms in a given situation (i.e., what we see as the descriptive norm). In many uncertain situations, we mimic the behavior of those around us because we think their behavior is normative. However, to natural resource managers, the mimicked behavior is often undesirable and might include a wide array of behaviors such as littering, approaching wildlife in parks, gathering firewood even if it is prohibited, ignoring caution signs. Assessing the offending group's norms might enhance our understanding of why people engage in these offensive behaviors. People may engage in offensive behavior because norms governing correct behavior are not evident or because the offending group does not believe the norms encouraging the correct behavior are salient. Cialdini et al. (1990) suggested that cues that focus people's attention on the norm, so that the norm becomes more salient, can affect people's littering behavior.

Predicting and Affecting Stakeholder Responses Natural resource management often involves prediction of human responses. Examples include predicting recreation participation, predicting response to educational or advertisement efforts, and predicting support for management actions.

Norms may be an important antecedent to human behavior; hence, they may also be important predictors of behavior. Research is now challenged to develop approaches that effectively measure the influence of norms on the individual. Several authors offer potential improvement by suggesting altering the Theory of Reasoned Action. Knowledge of normative influences is important because it can guide persuasion attempts.

Understanding Conflict Among Recreational Groups   The outdoor recreation literature notes repeatedly that conflict among stakeholders often emanates from groups or individuals adhering to different norms: Anglers may clash because of their differing norms about preferred distance from others (Martinson & Shelby, 1992); campers may clash because of norms about whether one should yell or whether or how loudly one should play music (Ruddell & Gramann, 1994); tourists may arouse local residents' animosity toward them by abandoning the propriety norms the tourists followed at home (Brown, 1999).

Understanding the differences in norms provides clues about dealing with this norm-based conflict. In some cases, resolving the conflict might be as simple as communicating the difference in norms between the conflicting groups so the offending behavior can be avoided.

Conducting Natural Resources Policy Analysis   Stakeholder responses to natural resources policy debates involve more than the economic or utilitarian qualities of outcomes. Often it is something much more fundamental; it can be a validation, enhancement, or protection of one's very identity. To illustrate, the mid-1980s witnessed a significant decline in salmon populations and, as a result, fish and wildlife managers dramatically reduced the length of salmon seasons. This had a significant effect on charter boats' captains whose ranks had grown during times of greater abundance and who depended upon revenue from their salmon fishing clientele. As seasons diminished, many captains went out of business, despite the fact that alternative business opportunities (i.e., bottom fishing, whale watching) seemed viable. Why did not captains switch to these alternatives? Part of the reason can be attributed to the perceived loss of identity. Many captains were strongly attached to the identity of a salmon captain, which they perceived to carry far more prestige and respect than the captain who offered tourist-oriented whale watching or bottom fishing. Much of the animosity over this situation was directed at the fish and wildlife agencies that implemented the season-reducing regulations. This illustrates a point by Cheng, Kruger, and Daniels (2003): Given that natural resource issues can be so closely aligned to ones self-identity, it "is not surprising that reactions to natural resource policy and management proposals can be so intensely emotional" (p. 93). Several authors emphasize the importance of identity's role in affecting the formation, endurance, and effectiveness of groups in policy debates. Clayton (2003) suggested that one's *environmental identity* would be an important force (more than attitudes) in directing thoughts and behavior about natural resource issues. Cheng et al. (2003) also proposed that the integration of place and identity is a critical process in the formation of individuals and groups that engage in the natural resources policy process. That is, common assignments of meaning and personal identity to specific locations become the foundation for groups that engage in policy discourse. Clearly, policy analysis that examines stakeholders based on variables associated with identity (e.g., values, norms) would add considerable insight in gauging the impacts of policy alternatives.

Summary

- Norms are informal rules of behavior among a defined social group. Although it might be difficult to identify the purpose of norms in every case, they generally serve to attain cooperation among group members. When acting cooperatively, as opposed to selfishly, there is a more uniform, predictable, and sustainable distribution of positive outcomes among the group.
- Norms have sanctions that maintain their continuity, i.e., those who do not obey a norm receive a sanction from other group members. Responses to those sanctions are largely dependent on one's attachment to the group; the greater is the attachment, the more likely one is to respond (e.g., be shamed).
- Normative influence might occur in various ways. Norms might arise in the form of social pressure from others who want a person to behave a specific way; they might emanate from one's own attachment to a group and the desire to be like the ideal group member; they might occur via the observation of others in uncertain situations; or they may occur as some combination of these effects.
- A wide variety of norms govern our day-to-day behavior; norms have this influence because we identify ourselves as members of many different groups and assume many roles within groups; each role has distinct normative associations.
- Norms are dependent upon a situation. Norms are not blindly obeyed; instead, they help us interpret a situation and select an appropriate behavior.
- Norms are a human universal because they have a selective advantage. Norms have a selective advantage because following them results in cooperative behavior.
- A backward-looking view suggests norms exist because they are regularities in behavior that sustain group solidarity. A forward view suggests norms are adopted because they offer benefits to those who adhere to them.
- Frequently used theoretical approaches to norms include Schwartz's Theory of Moral Decisions and his Norm Activation Model, Fishbein and Ajzen's Theory of Reasoned Action, Norm Focus Theory, Identity Theory, Self-Identity Theory and Self-Categorization Theory.
- A framework that integrates topics introduced in this book – attitudes, values, norms, and emotions – would be useful in exploring groups involved in fish and wildlife issues.

References

Aberle, D. F., Cohen, A. K., Davis, A. K., Levy, M. J., & Sutton, F. X. (1950). The functional prerequisites of a society. *Ethics, 60*(2), 100–111.
Abrams, D., Ando, K., & Hinkle, S. (1998). Psychological attachment to the group: Cross cultural differences in organizational identification and subjective norms as predictors of

workers' turnover intentions. *Personality and Social Psychology Bulletin, 24*(10), 1027–1039.

Ajzen, I. (1991). Theory of planned behavior. *Organizational Behavior and Human Decision Processes, 50*, 170–211.

Ajzen, I., & Fishbein, M. (1980). *Understanding attitudes and predicting social behavior.* Englewood Cliffs, NJ: Prentice-Hall.

Armitage, C. J., & Conner, M. (2001). Efficacy of the theory of planned behaviour: A meta-analytic review. *British Journal of Social Psychology, 40*, 471–499.

Astrom, A., & Rise, J. (2001). Young adults' intention to eat healthy food: Extending the theory of planned behavior. *Psychology and Health, 16*(2), 223–237.

Barro, S. C., & Manfredo, M. J. (1996). Constraints, psychological investment, and hunting participation: Development and testing of a model. *Human Dimensions of Wildlife, 1*(3), 42–61.

Biddle, B., Bank, B., & Slavings, R. (1987). Norms, preferences, identities and retention decisions. *Social Psychology Quarterly, 50*(4), 322–337.

Bratt, C. (1999). The impact of norms and assumed consequences on recycling behavior. *Environment and Behavior, 31*(5), 630–656.

Brewer, M. B. (1999). The psychology of prejudice: Ingroup love or outgroup hate? *Journal of Social Issues, 55*(3), 429–444.

Brown, T. J. (1999). Antecedents of culturally significant tourist behavior. *Annals of Tourism Research, 26*(3), 676–700.

Charng, H. W., Piliavin, J. A., & Callero, P. L. (1988). Role identity and reasoned action in the prediction of repeated behavior. *Social Psychology Quarterly, 51*(4), 303–317.

Cheng, A. S., Kruger, L. E., & Daniels, S. E. (2003). "Place" as an integrating concept in natural resource politics: Propositions for a social science research agenda. *Society and Natural Resources, 16*, 87–104.

Christian, J., & Armitage, C. J. (2002). Attitudes and intentions of homeless persons toward service provision in South Wales. *The British Journal of Social Psychology, 41*(2), 219–231.

Christian, J., & Abrams, D. (2003). The effects of social identification, norms and attitudes on use of outreach services by homeless people. *Journal of Community and Applied Social Psychology, 13*(2), 138–157.

Christensen, P. N., Rothgerber, H., Wood, W., & Matz, D. C. (2004). Social norms and identity relevance: A motivational approach to normative behavior. *Personality and Social Psychology Bulletin, 30*(10), 1295–1309.

Christian, J., Armitage, C. J., & Abrams, D. (2003). Predicting uptake of housing services: The role of self categorization in the theory of planned behaviour. *Current Psychology, 22*(3), 206–217.

Cialdini, R. B., Reno, R. R., & Kallgren, C. A. (1990). A focus theory of normative conduct: Recycling the concept of norms to reduce littering in public places. *Journal of Personality and Social Psychology, 58*, 1015–1026.

Cialdini, R. B., & Trost, M. R. (1998). Social influence: Social norms, conformity, and compliance. In D.T. Gilbert, & S.T. Fiske (Eds.), *The handbook of social psychology*, (4th ed., pp. 151–192). Boston: McGraw-Hill.

Clayton, S. (2003). Environmental identity: A conceptual and operational definition. In S. Clayton, & S. Opotow (Eds.), *Identity and natural environment: The psychological significance of nature* (pp. 45–65). Cambridge: The MIT Press.

Coleman, J. S. (1990). *Foundations of social theory.* Cambridge, MA: Harvard University Press.

Conner, M., & McMillan, B. (1999). Interaction effects in the theory of planned behaviour: Studying cannabis use. *British Journal of Social Psychology, 38*(2), 195–222.

Darley, J. M., & Latane, B. (1970). Norms and normative behavior: Field studies of social interdependence. In J. Macaulay, & L. Berkowitz (Eds.), *Altruism and helping behavior* (pp. 83–102). New York: Academic Press.

Deaux, K. (1996). Social identification. In E. T. Higgins, & A. W. Kruglanski (Eds.), *Social psychology: Handbook of basic principles* (pp. 777–797). New York: Guilford Press.

DeVries, H., Backbier, E., Kok, G., & Dijkstra, M. (1995). The impact of social influences in the context of attitude, self efficacy, intention, and previous behavior as predictors of smoking onset. *Journal of Applied Social Psychology, 25*(3), 237–257.

Donnelly, M. P., Vaske, J. J., Whitaker, D., & Shelby, B. (2000). Toward an understanding of norm prevalence: A comparative analysis of 20 years of research. *Environmental Management, 25*(4), 403–414.

Elster, J. (1989). *The cement of society: A study of social order*. Cambridge, MA: Cambridge University Press.

Elster, J. (2003). Coleman on social norms. *Revue Francaise de Sociologie, 44*(2), 297–304.

Eriksen, T. H. (1995). *Small places, large issues: An introduction to social and cultural anthropology* (2 nd ed.). London: Pluto Press.

Farley, J. U., Lehmann, D. R., & Ryan, M. J. (1981). Generalizing from "imperfect" replication. *Journal of Business, 54*(4), 597–610.

Fehr, E., & Fischbacher, U. (2004). Social norms and human cooperation. *Trends in Cognitive Sciences, 8*(4), 185–190.

Fekadu, Z., & Kraft, P. (2002). Expanding the theory of planned behavior: The role of social norms and group identification. *Journal of Health Psychology, 7*(1), 33–43.

Fine, G. A. (2001). Enacting norms: mushrooming and the culture of expectations and explanations. In M. Hecter, & K. D. Opp (Eds.), *Social Norms* (pp. 139–164). New York: Russell Sage Foundation.

Fishbein, M., & Ajzen, I. (1975). *Belief, attitude, intention, and behavior: An introduction to theory and research*. Reading, MA: Addison-Wesley.

Forgas, J. P., & Williams, K. D. (2001). Social influence: Introduction and overview. In J. P. Forgas, & K. D. Williams (Eds.), *Social Influence: Direct and indirect processes* (pp. 3–24). Philadelphia, PA: Psychology Press.

Fulton, D. C., Pate, J., & Manfredo, M. J. (1995). *Colorado resident's attitudes toward trapping in Colorado*. (Project Report No. 23). Project report for the Colorado Division of Wildlife. Fort Collins: Colorado State University, Human Dimensions in Natural Resources Unit.

Gintis, H. (2003). Solving the puzzle of prosociality. *Rationality and Society, 15*(2), 155–187.

Granberg, D., & Holmberg, S. (1990). The intention-behavior relationship among U.S. and Swedish voters. *Social Psychology Quarterly, 53*, 44–54.

Grube, J. W., Morgan, M., & McGree, S. T. (1986). Attitudes and normative beliefs as predictors of smoking intentions and behaviours: A test of three models. *British Journal of Social Psychology, 25*, 81–93.

Hall, T. E., & Roggenbuck, J. W. (2002). Response format effects in questions about norms: Implications for the reliability and validity of the normative approach. *Leisure Sciences, 24*, 325–337.

Heberlein, T. A. (1972). The land ethic realized: Some social psychological explanations for changing environmental attitudes. *Journal of Social Issues, 28*(4), 79–87.

Heberlein, T. (1977). Density, crowding, and satisfaction: Sociological studies for determining carrying capacities. *Proceedings: River Recreation Management and Research Symposium*. USDA Forest Service General Technical report NC-28, 67–76.

Hechter, M. (1987). *Principles of group solidarity*. Berkley, CA: University of California Press.

Hecter, M., & Opp, K. D. (2001). What have we learned about the emergence of social norms? In M. Hecter, & K. D. Opp (Eds.), *Social Norms* (pp. 394–415). New York: Russell Sage Foundation.

Heywood, J. L. (1996). Convention, emerging norms, and norms in outdoor recreation. *Leisure Sciences, 18*, 355–363.

Heywood, J. L. (2000). Current approaches to norm research. In D. N. Cole, S. F. McCool, W. T. Borie, & J. O'Loughlin (Comps.), *Wilderness science in a time of change – Volume 4:*

Wilderness visitors, experiences, and visitor management (pp. 260–264). Proceedings RMRS-P-15-Vol-4, Odgen, UT: U.S. Dept. of Agriculture, Forest Service, Rocky Mountain Research Station.

Hopper, J. R., & Nielsen, J. M. (1991). Recycling as altruistic behavior: Normative and behavioral strategies to expand participation in a community recycling program. *Environment and Behavior, 23*(2), 195–220.

Horne, C. (2001a). Sociological perspectives on the emergence of norms. In M. Hecter, & K. D. Opp (Eds.), *Social norms* (pp. 3–34). New York: Russell Sage Foundation.

Horne, C. (2001b). Sex and sanctioning: Evaluating two theories of norm emergence. In M. Hecter, & K. D. Opp (Eds.), *Social norms* (pp. 305–324). New York: Russell Sage Foundation.

Hrubes, D., Ajzen, I., & Daigle, J. (2001). Predicting hunting intentions and behavior: An application of the theory of planned behavior. *Leisure Sciences, 23*, 165–178.

Jackson, J. (1965). Structural characteristics of norms. In I. D. Steiner, & M. J. Fishbein (Eds.), *Current studies in social psychology* (pp. 301–309). New York: Holt, Rinehart and Winston.

Jetten, J., Postmes, T., & McAuliffe, B. J. (2002). 'We're all individuals': Group norms of individualism and collectivism, levels of identification and identity threat. *European Journal of Social Psychology, 32*(2), 189–207.

Kenrick, D., Ackerman, J., & Ledlow, S. (2003). Evolutionary social psychology: Adaptive predispositions and human culture. In J. Delamater (Ed.), *Handbook of social psychology* (pp. 103–122). New York: Kluwer.

Krebs, D. L., & Miller, D. T. (1985). Altruism and aggression. In G. Lindzey, & E. Aronson (Eds.), *The handbook of social psychology* (3rd ed., Vol. 2., pp. 1–71). New York: Random House.

Manning, R. E. (2007). *Parks and carrying capacity: Commons without tragedy*. Washington, DC: Island Press.

McCall, G., & Simmons, J. L. (1978). *Identities and interactions: An examination of human associations in everyday life* (Revised edition). New York: The Free Press.

Martinson, K. S., & Shelby, B. (1992). Encounter and proximity norms for salmon anglers in California and New Zealand. *North American Journal of Fisheries Management, 12*(3), 559–567.

Miller, D. T., & Prentice, D. A. (1996). The construction of social norms and standards. In E. T. Higgins, & A. W. Kruglanski (Eds.), *Social psychology: Handbook of basic principles* (pp. 799–829). New York: Guilford Press.

Opp, K. D. (2001). Social networks and the emergence of protest norms. In M. Hecter, & K. D. Opp (Eds.), *Social norms* (pp. 234–273). New York: Russell Sage Foundation.

Monroe, M., Newman, P., Pilcher, E., Manning, R., & Stack, 10. (2007). Now hear this. *Legacy, 18*(1), 18–25.

Robbins, J. M., & Krueger, J. I. (2005). Social projection to ingroups and outgroups: A review and meta-analysis. *Personality and Social Psychology Review, 9*(1), 32–47.

Roggenbuck, J. W., Williams, D. R., Bange, S. P., & Dean, D.J. (1991). River float trip encounter norms: Questioning the use of the social norms concept. *Journal of Leisure Research, 23*(2), 133–153.

Rosenberg, M. (1979). *Conceiving the self*. New York: Basic Books.

Ruddell, E. J., & Gramann, J. H. (1994). Goal orientation, norms and noise-induced conflict among recreation area users. *Leisure Sciences, 16*(2), 93–104.

Schwartz, S. H. (1968). Awareness of consequences and the influence of moral norms on interpersonal behavior. *Sociometry, 31*(4), 355–369.

Schwartz, S. H. (1977). Normative influences on altruism. In L. Berkoowitz (Ed.), *Advances in experimental social psychology* (Vol. 10, pp. 221–279). New York: Academic Press.

Schwartz, S. H., & Howard, J. A. (1982). Helping and cooperation: A self-based motivational model. In V. J. Derlega, & J. Grezlak (Eds.), *Cooperation and helping behavior: Theories and Research* (pp. 327–353). New York: Academic Press.

Schofield, P. E., Pattison, P. E., Hill, D. J., & Borland, R. (2001). The influence of group identification on the adoption of peer group smoking norms. *Psychology and Health, 16*, 1–16.

Shelby, B. (1981). Encounter norms in backcountry settings: Studies of three rivers. *Journal of Leisure Research, 13*, 129–138.

Shelby, B., & Heberlein, T. A. (1986). *Carrying capacity in recreation settings*. Corvallis: Oregon State University Press.

Shelby, B., & Vaske, J. J. (1991). Using normative data to develop evaluative standards for resource management: A comment on three recent papers. *Journal of Leisure Research, 23*, 173–187.

Shelby, B., Vaske, J. J., & Donnelly, M. P. (1996). Norms, standards, and natural resources. *Leisure Sciences, 18*, 103–123.

Shepherd, G. J., & O'Keefe, D. J. (1984). Separability of attitudinal and normative influences on behavioral intentions in the Fishbein-Ajzen model. *Journal of Social Psychology, 122*, 287–288.

Stryker, S. (1980). *Symbolic interactionism: A socio-structural version*. Menlo Park, CA: Benjamin Cummings.

Tajfel, H. (1981). *Human groups and social categories: Studies in social psychology*. England: Cambridge University Press.

Terry, D. J., & Hogg, M. A. (1996). Group norms and attitude-behavior relationship: A role for group identification. *Personality and Social Psychology Bulletin, 22*(8), 776–793.

Terry, D. J., Hogg, M. A., & White, K. M. (1999). The theory of planned behaviour: Self-identity, social identity, and group norms. *British Journal of Social Psychology, 38*(3), 225–244.

Terry, D. J., Hogg, M. A., & White, K. M. (2000). Attitude-behavior relations: Social identity and group membership. In D. J. Terry, & M. A. Hogg (Eds.), *Attitudes, behavior, and social context: The role of norms and group membership* (pp. 67–93). Mahwah, NJ: Lawrence Erlbaum Associates.

Theodorakis, Y. (1994). Planned behavior, attitude strength, role identity and the prediction of exercise behavior. *The Sport Psychologist, 8*, 149–165.

Trafimow, D., & Finlay, K. (1996). The importance of subjective norms for a minority of people: Between-participants and within participants analyses. *Personality and Social Psychology Bulletin, 22*, 820–828.

Trafimow, D., & Fishbein, M. (1994). The moderating effect of behavior type on the subjective norm-behavior relationship. *Journal of Social Psychology, 134*(6), 755–763.

Turner, R. H. (1978). The role and the person. *American Journal of Sociology, 84*, 1–23.

Turner, J. C., Hogg, M. A., Oakes, P. J., Reicher, S. D., & Wetherell, M. S. (1987). *Rediscovering the social group: A self-categorization theory*. Oxford: Basil Blackwell.

Vallerand, R. J., Pelletier, L. G., Deshaies, P., Cuerrier, J. P., & Mongeau, C. (1992). Ajzen and Fishbein's theory of reasoned action as applied to moral behavior: A confirmatory analysis. *Journal of Personality and Social Psychology, 62*, 98–109.

Vaske, J. J, Shelby, B. B., Graef, A. R., & Heberlein, T. A. (1986). Backcountry encounter norms: Theory, method and empirical evidence. *Journal of Leisure Research, 18*, 137–153.

Vaske, J. J., & Whittaker, D. (2004). Normative approaches to natural resources. In M. J. Manfredo, J. J. Vaske, B. L. Bruyere, D. R. Field, & P. J. Brown (Eds.), *Society and Natural Resources: A summary of knowledge prepared for the 10th International Symposium on Society and Natural Resources* (pp. 283–294). Jefferson, MO: Modern Litho.

Wagar, A. J. (1964). *The carrying capacity of wild lands for recreation*. Forest Science Monograph, 7. Washington, DC: Society of American Foresters.

Williams, D. R., Roggenbuck, J. W., Patterson, M. E., & Watson, A. E. (1992). The variability of user-based social impact standards for wilderness management. *Forest Science, 38*(4), 738–756.

Ybarra, O., & Trafimow, D. (1998). How priming the private self or collective self affects relative weights of attitudes and subjective norms. *Personality and Social Psychology Bulletin, 24*(4), 362–370.

Young, R. A., & Kent, A. T. (1985). Using the theory of reasoned action to improve the understanding of recreation behavior. *Journal of Leisure Research, 17*(2), 90–106.

Zinn, H. C., Manfredo, M. J., Vaske, J. J., & Wittmann, K. (1998). Using normative beliefs to determine the acceptability of wildlife management actions. *Society and Natural Resources, 11*, 649–662.

Chapter 6
Values, Ideology, and Value Orientations

Contents

Introduction

Think about an experience that you had with wildlife that made you happy. How would you describe that experience? Would it, for example, involve watching animals, hunting for animals, saving or protecting animals, or showing affection to animals? How was the animal treated? Was the animal an object for human use, or was it depicted as a potential companion? Did the animal arouse affection? Do you see the animal through the eyes of science, or does your story suggest religious or spiritual meaning?

Each story a person tells us when asked this question indicates the individual's thought processes, and we have used this story-telling approach to detect people's wildlife value orientations in a recent global study (see Dayer, Stinchfield, & Manfredo, 2007). The stories people tell indicate something about their values.

Values are critical because they (a) represent an individual's personal goals and standards for determining good and bad or right and wrong, (b) guide a person in interpreting events and information, and (c) are present across situations and events.

M.J. Manfredo, *Who Cares About Wildlife?*,   141
DOI: 10.1007/978-0-387-77040-6_6, © Springer Science+Business Media, LLC 2008

From the social science view, values have important implications. If we understand a person's values toward wildlife, we understand how that person will think and behave in wildlife-associated situations. In this chapter, I review key theories about values, ideology, and value orientations and suggest how these theories help us understand human–wildlife relationships.

Origins of Interest in the Wildlife Values Topic

Scientific attention given to the topic of wildlife values has grown considerably since the latter third of the twentieth century. This interest was stimulated largely by the practical concerns of the wildlife management profession. One of these concerns was how to show the worth of wildlife to an apparently oblivious society.

In introducing the 1987 volume *Valuing Wildlife* (Decker & Goeff, 1987), Robert Chambers lauded the vision of the early leaders of the wildlife profession – Leopold, Errington, King, and Stoddard – as recognizing "our need to identify and project [wildlife] values to a society and a world driven by forces of cash income, profit motive, and a soaring technology, which were eliminating wildlife and its habitat at a prodigious rate through deforestation, intensified agriculture, wetland drainage, development and pollution" (p. xvii).

Early work on the topic of wildlife values readily embraced this mission. Noted by King (1947), "Can it be determined that these [wildlife values] are sufficient to justify costs in time, labor, land and money necessary for the conservation and management of this resource?" (p. 456).

In the 1970s, researchers began to use established social science methods to study wildlife values. As this research unfolded, the prediction of the early leaders of the wildlife profession was fulfilled. Results of this research documented a wide array of economic and social benefits derived from the presence and recreational enjoyment of wildlife. Research findings continue to expand our understanding of wildlife values as studies take an international scope and embrace an ever-widening array of issues and concerns; however, a caution given by Bryan (1980) still applies today. Bryan (1980), who presented at a 1979 workshop titled *Wildlife Values*, noted that values and motivation research about wildlife then being conducted lacked a sound conceptual foundation, which "without an attempt to integrate data into a coherent scheme leaves us without a conceptual 'map' to guide and interpret future research" (p. 72).

This chapter overviews the conceptual approaches used for examining human values generally and human wildlife value orientations more specifically.

As a preface, note the distinction between the term *values* used as a verb and the term *values* used as a noun (Rohan, 2000). As a verb, *values* focuses on people's assignment of meaning, goodness or worth. Natural resource economists, for example, use willingness-to-pay methods to determine the dollar value of a particular wildlife species. Similarly, a social psychologist might determine a person's positive or negative evaluation of an issue (see Chapter 4). As a noun, a *value* is "a stable, meaning-producing, super-ordinate cognitive structure" (Rohan, 2000, p.257). These two uses are dependant concepts. The process of valuing (verb) an object emanates from the enduring values (noun) that a person holds.

This chapter focuses on the enduring cognitions that affect our thought about wildlife, i.e., on the use of *values* as a noun. The chapter begins by reviewing earlier wildlife values research. It then reviews theory regarding values, ideology, and value orientations. The chapter ends focusing on a theory of wildlife value orientations.

Prior Research on Wildlife Values

In early research on values toward wildlife, attitudes toward wildlife, and motivations for recreation associated with wildlife, different researchers have taken similar paths (see, for example, Potter, Hendee, & Clark, 1973; Hendee, 1974; Hautaluoma & Brown, 1978; Purdy & Decker, 1989). The prevailing approach is empirical, not theoretical (Manfredo, Teel, & Bright, 2004).

What is the difference between an empirical approach and a theoretical approach? In the empirical approach, researchers collect data, organize it, and develop post hoc explanations for patterns they find. With a theoretical approach, researchers construct an explanation (by building upon prior scientific work), develop measures that represent concepts in the explanation, collect data, and test the explanation.

Following the empirical approach, early wildlife values research began with interviewing a sample of people who represented the population of interest. Researchers obtained statements that reflected the concept of interest by asking interviewees questions like, "For what reasons do you participate in hunting?," "Why is wildlife important to you?," and "How do you view our relationship with other animals?" From the answers interviewees provided, the researcher developed fixed-response survey items. Typically, multiple survey items represented a common theme found in participants' responses; each theme was interpreted as a *type* of value or motivation. The surveys developed using this method were then administered to a large sample of participants, and the results were analyzed using factor or cluster analysis techniques. The resultant groupings were labeled and taken to represent the value typology for the concept of interest. With this approach, theoretical explanations, if offered at all, were developed to fit the empirical findings. The item groupings are sometimes even referred to as the investigator's theory of values.

A number of studies empirically describe wildlife values. Decker and Geoff (1987) and Shaw and Zube (1980) overviewed and summarized various early approaches.

The next section reviews a popular approach that has received the widest attention, Kellert's *typology of wildlife attitudes* (Kellert, 1976; Kellert, 1993; Kellert, 2002).

Kellert's typology. Kellert (1976) developed his approach when interest in the social aspects of wildlife was emerging and little scientific information about the topic existed. He presented a typology of "attitudes toward wildlife" (he also referred to them as values) that has remained the central thrust of his writings.

Kellert constructed his typology using the approach described above. He interviewed 65 people who had various wildlife-related interests. From these interviews, he considered and pretested over 1,000 survey items; of those 1,000 items, he selected 79 to include in his data collection instrument. These 79 items were grouped so that they represented the nine separate values shown in Table 6.1.

Subsequent work applies Kellert's typology descriptively. For example, in a study intended to describe the American public, Kellert (1980) found that that the *defining attitudes* among the American public during the 1970s were humanistic attitudes (35% strongly oriented toward this attitude), moralistic (20%), utilitarian (20%), and neutralistic (35%). Studying attitudes toward predators, he found positive responses to predators were related to positive naturalistic, moralistic, and ecologistic attitudes, but negatively related to negativistic and utilitarian scales (Kellert, 1985). While exploring gender differences, Kellert and Berry (1987) indicated that females have higher scores than males on humanistic, moralistic, and negativistic values. Men were higher on utilitarian, dominionistic, naturalistic, and ecologistic attitudes than were females. While researching hunters and antihunters, Kellert (1978) found that *meat hunters* were characterized as having utilitarian attitudes, *nature hunters* were high on naturalistic attitudes and *sport hunters* had strong dominionistic attitudes. In contrast, antihunters were grouped into two types: the *humanistic antihunter* and the *moralistic antihunter*.

Table 6.1 Kellert's typology of basic values

Value	Definition
Utilitarian	Practical and material exploitation of nature
Naturalistic	Direct experience and exploration of nature
Ecologistic-scientific	Systematic study of the structure, function, and relationship in nature
Aesthetic	Physical appeal and beauty of nature
Symbolic	Use of nature for language and thought
Humanistic	Strong emotional attachment and "love" for aspects of nature
Moralistic	Spiritual reverence and ethical concern for nature
Dominionistic	Mastery, physical control, and dominance of nature
Negativistic	Fear, aversion, and alienation from nature

(Source: Kellert, 1996).

Kellert and others emphasized the importance of considering the information obtained via his typology of attitudes in the decision-making context (e.g., Langenau, Kellert, & Applegate, 1984). He suggested, for example, that when conducting cost–benefit analysis, policy makers consider more than just dollar measures. The typology of attitudes captures the diversity of ways people view and appreciate wildlife (Kellert, 1984). More broadly, Clark and Kellert (1988) proposed a policy paradigm for the wildlife sciences that explicitly recognizes valuation (including Kellert's value typology) along with biophysical, authority/property, and institutional decision-making variables.

In the 1970s and 1980s, Kellert identified his typology of attitudes with a social psychological framework (e.g., Kellert, 1983). He suggested the importance of understanding affective, cognitive, and evaluative perceptions about wildlife. However, his model did not propose any relationships among these concepts or between these attitudes and other conceptual or behavioral domains.

In the early 1990s, Kellert's orientation shifted; biophilia became the basis of his attitude typology (Kellert & Wilson, 1993; Kellert, 2000). Biophilia proposes that humans are innately inclined to affiliate with other living things and that this inclination has offered our species a competitive advantage throughout our evolutionary history. Kellert proposed that his typology of attitudes represents nine different "expressions of the biophilia tendency" (1993, p. 43). As he conceptually transitioned from a social psychological to a sociobiology orientation, Kellert kept his attitude descriptions unchanged, and no empirical evidence yet supports the conceptual shift.

Strengths and limitations of Kellert's work. Kellert's work contributes to our understanding of people's values toward wildlife in three ways. First, his work was one of the first to use a social science approach to understand wildlife values. This step showed the overall relevance of the social sciences to wildlife decision making. Second, Kellert's research describes the various ways people consider wildlife and, importantly, how opposing values can be the basis for conflict among different groups of people. Kellert (1980) was also one of the first to suggest, based on empirical findings, that a shift in wildlife values was occurring in the United States. Third, his research effects how wildlife professionals and scientists view human relations with animals. His work encourages wildlife professionals to reject the assumption that their own personal views toward wildlife mirrored the public's views about wildlife. In fact, people hold a diversity of attitudes and beliefs about wildlife. Descriptions from Kellert's typology are still present in the lexicon of some wildlife professionals.

We also must recognize both the methodological and conceptual limitations in Kellert's approach. A prominent concern relates to the methodological adequacy of the instrument used to measure Kellert's nine attitude types. Though Kellert acknowledged the complexity of psychometric scales (e.g., Kellert, 1980), very little has been published that addresses issues of reliability and validity regarding the use of his scales. This lack of evidence is undoubtedly due to the fact that Kellert has not published his scales in a widely accessible location; hence, other researchers rarely use the scales (e.g., Kaltenborn & Bjerke, 2002). Among these studies, Vitterso, Bjerke, and Kaltenborn (1999) reported alpha reliability indices for some item groupings (used to measure an attitude type) in the 0.50–0.69 range, which is below the 0.70 threshold recommended by standards psychometric texts (e.g., Nunnally & Bernstein, 1994). Vitterso et al. (1999) also found a good fit for a model that grouped Kellert's scales into two basic categories which they labeled *positive attitudes* and *negative attitudes*.

Such findings are troubling, but inconclusive. However, the general paucity of tests of the Kellert method raises the possibility that there may be (a) a more appropriate structure underlying the data collected with Kellert's scales or (b) that the scale structure (as evidenced, for example, by confirmatory factor analysis) does not hold for all sample types.

The second concern involves the lack of a clear conceptual foundation for the psychological concepts being measured with Kellert's instrument. Kellert maintained that his scaling measured basic attitudes toward animals and alternatively referred to them as values, attitudes, perceptions, and evaluations. More recently, Kellert referred to them as inherited tendencies.

The lack of a conceptual orientation results in contradictory inferences. Can these attitudes be changed (as attitude theory would suggest), or are they set in childhood and endure through one's lifetime (as value theory would suggest)? How does one come to acquire certain attitudes? Are they learned (as value theory suggests) or is one born with those dispositions (as biophilia suggest)? If they are inherited traits, why is there such variability among the nature of these tendencies? Finally, are these attitudes supposed to predict behaviors, and if so, which behaviors do they predict?

In summary, Kellert's approach, like many approaches from the 1970s and 1980s, contributes to the description of wildlife values; however, future work on wildlife values must advance beyond these conceptual and methodological limitations.

The next section provides an overview of the prominent theoretical approaches to social values and clarifies key characteristics of the values concept. A theory of wildlife value orientations is offered in the chapter's conclusion.

Theory on Social Values

Psychology, sociology, and anthropology literature reference and describe values regularly. This widespread use complicates the term by assigning it many contrasting definitions (Kluckhohn, 1951; Campbell, 1963; Rohan, 2000). This book's scope is too limited to review these many definitions; however, it will touch upon the two most widely referenced approaches. One of these approaches has been applied to human–wildlife relationships.

Among contemporary researchers, Milton Rokeach was one of the most important in stimulating work on the values concept during the latter half of the twentieth century. Allport's, Kluckholn's, and William's work guided and shaped Rokeach's concepts. Allport (1961) saw value priorities as the dominating forces in a person's life; Kluckholn (1951) saw value orientations as conceptions "of nature, man's place in it, man's relations to man, and of the desirable and undesired within man–nature and man–man relations" (p. 411), and Williams (1968) suggested that values are the criteria used when people make evaluations.

Elements of these three approaches can be seen in Rokeach's definition of values, which is "an enduring belief that a specific mode of conduct or end state of existence is personally or socially preferable to an opposite or converse mode of conduct or end state of existence" (1973, p. 5). Rokeach's definition emphasizes that a value is a belief that includes a cognitive component (an aspect that is true or false to the holder), an affective component (including elements of emotion and evaluation of liking or disliking), and a behavioral component (linkage to certain classes of actions). Moreover, they are taught as absolute truths in our youth and are enduring throughout life.

Rokeach proposed that values are organized into a *value system*. A value system is an enduring organization of beliefs or values about modes of conduct or end states of existence. He referred to values regarding modes of conduct as *instrumental values* and values addressing end states of existence as *terminal values*. Among terminal values, he proposed there were personal values (focused on the individual's well-being) and social values (focused on society at large). Similarly, instrumental values dealt with what is moral or with one's competence.

Rokeach proposed that value systems serve several important functions. First and foremost, value systems provide standards that guide our activities. They are ubiquitous in our lives: they lead us to take a particular position on social issues, they predispose us to a particular religious or political ideology, they guide our presentations of ourselves to others, and they cause us to evaluate others and rationalize our own otherwise unacceptable actions. Second, value systems guide our conflict resolution and decision making. Because conflicts typically involve competing values, priorities we assign to values can guide resolutions. Finally, values function to give expression to basic human needs. Rokeach proposed that terminal values represent "supergoals beyond immediate, biologically urgent goals" (1973, p. 14).

Noting the preponderance of attitude studies in psychology, Rokeach carefully differentiated *values* from *attitudes*. He proposed that (a) values transcend objects and situations, while attitudes focus on specific objects or situations, (b) values are single beliefs, while attitudes organize several beliefs around an object, and (c) values occupy a more central position than attitudes and "are therefore determinants of attitudes" (1973, p. 18).

While most people in a society share a certain number of values, individuals are unique because of the prioritization of values, or value systems, that they hold. Rokeach contended that these individual differences arise in the personal, societal, and cultural experiences. Within this theoretical context, Rokeach developed a typology of 18 instrumental and 18 terminal values. His theoretical and methodological approach, although widely used and referenced (Rohan, 2000), is slowly being replaced by Schwartz's approach to values.

Schwartz's value theory. Shalom Schwartz's theory advances conceptions of value structures and the implications these structures have on understanding differences in human behavior (Schwartz, 1992, 1996, 2006). Schwartz suggested that values represent what is important in our lives and that values serve as goals that apply across contexts and time. Values are stable motivational constructs that change little in a person's adult life. Schwartz suggested there is a universal set of value types and that people's value structures differ because they prioritize these value types differently. Schwartz viewed values from a functional perspective, i.e., they allow humans to adapt to their surroundings. In that regard, they guide activities that lead to fulfillment of needs related to biological requisites, social interaction, and group survival and functioning.

Schwartz proposed a typology of ten values. These values represent two bipolar motivational dimensions (see Fig. 6.1). The arrangement of values into a circle presents an important characteristic of the values: conflicting values are in opposite directions from the center while values adjacent to one another are congruent. For example, activities that lead to fulfillment of the power value conflict with attainment of the universalism value (which Schwartz associated with environmentalism). Also depicted in this figure are the two main motivational constructs that underlie the values. One dimension is referred to as *openness to change and conservation* while the other is *self-enhancement – self-transcendence*. The motivational dimensions are aligned on the circle with the values with which they associate. For example, values focused on power and achievement are reflections of motivation for self-enhancement while values focused on universalism and benevolence are associated with self-transcendence.

Schwartz contended that values direct human behavior and attitudes, i.e., there is consistency between people's values and their overt behavior. When studies examine *value relevant* behaviors, moderate prediction occurs from value priority scoring (Bardi & Schwartz, 2003; Feather, 1988; Schwartz, 1996). In the cross-cultural context, results show moderate prediction in the area of religion, politics, and social relations (Smith & Schwartz, 1997).

The theoretical structure proposed by Schwartz has received strong cross-cultural validation. Working with collaborators in 54 countries, data has been gathered on 44,000 respondents. Repeated multidimensional scaling analyses provide "substantial support for the near universality of the ten value types and their structural relations" (Smith & Schwartz, 1997, p. 88). Sturch, Schwartz, and Kloot (2002), for example, showed that the proposed underlying value structure of Schwartz's scales does not vary across eight cultural regions, by gender, or by the interaction of cultural region and gender.

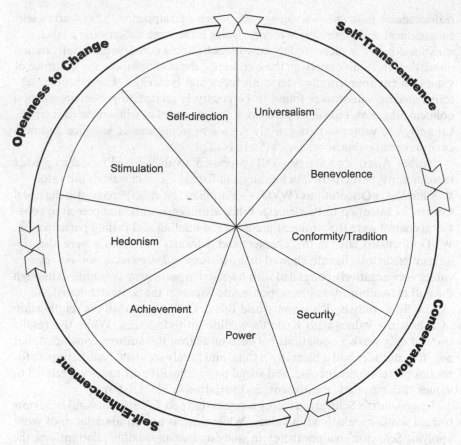

Fig. 6.1 Schwartz's values typology (Source shown below). Adapted from Schwartz (1992), with kind permission from Elsevier
Source: Schwartz, S.H. (1992). Universals in the content and structure of values: Theoretical advances and empirical tests in 20 countries. In M. Zanna (Ed.), *Advances in experimental social psychology* (Vol. 25, p. 1-65). New York: Academic Press.

Schwartz's theory has been used successfully to examine environmental issues and human–wildlife issues. For example, the work of Dietz, Guagnano, Kalof, and Stern (Stern, Dietz, Kalof, & Guagnano, 1995; Stern, Dietz, & Kalof, 1993; Stern, Dietz, & Guagnano, 1995; Dietz, Frisch, Kalof, Stern, & Guagnano, 1995; Stern & Dietz, 1994) has applied Schwartz's approach in studies that deal with environmentalism-related topics. These researchers originally proposed a classification of three environmentally related value orientations: egoistic (self-interest), social altruistic (concern for welfare of other humans), and biospheric (concern for non-human species or the biosphere). In later work, they explored the overlap of their own value typology with that proposed by Schwartz and developed a hybrid of the two models (Stern, Dietz, & Guagnano, 1995, 1998). The merged approach used one factor they labeled *biospheric–altruistic* that overlapped with Schwartz's transcendence motivation scale; *egoistic,* which overlapped with Schwartz's self-enhancement scale; *openness to change,* which is the same as Schwartz's label, and *traditional values,* which overlapped with Schwartz's conservation motivational dimension. In studies applying these concepts, the authors show the influence of values on environmentally related attitudes and behaviors. For example, self-transcendence values were found to be positively related to pro-environmental political behaviors, consumer behaviors, and measures of willingness to sacrifice. Conservation values were negatively related to willingness to sacrifice and pro-environmental political behaviors (Stern et al., 1998).

Hrubes, Ajzen, and Daigle (2001) proposed a values-based model to predict recreationists' participation in hunting and fishing, i.e., Schwartz's life values → Wildlife Value Orientations (WVO) → Attitudes, Norms, Perceived Behavioral Control → Intention to Participate. While attitudes, norms, and perceived behavioral control were the strongest predictors of hunting and fishing participation, WVO (discussed later in this chapter) and Schwartz's life values were also significant predictors. Results showed that *self-transcendence* and *openness to change* values were negatively correlated with *intention to participate in hunting* (although the authors cautioned on low response rate issues on the Schwartz items).

Finally, a study by Kaltenborn and Bjerke (2002) examined the relationship of Schwartz's values with Kellert's wildlife attitude scales. While the results showed only weak associations between measures, the authors concluded that negative attitudes had a basis in personal and family security, health, respect for traditions, economic income, and social power. Positive attitudes are related to values "like curiosity, excitement, and variation in life" (p. 60).

To summarize, Schwartz's theory suggests that people's attitudes and behaviors toward wildlife are rooted in values. While there is not an abundance of work applying Schwartz's values theory in studying human–wildlife relationships, the research that is available suggests people with strong *conservation* and *self-enhancement* values will be more likely to hold utilitarian and domination views of wildlife while those with strong *openness to change* and *self-transcendence* values will be more likely to hold aesthetic and mutualistic views toward wildlife. Future research is needed to confirm or reject this hypothesis.

Characteristics of the Values Concept

Rockeach and Schwartz represent social psychology's predominant theoretical approaches to values; however, in addition to the social psychology approach, value theory has many facets that can be used in the human–wildlife field. This section identifies several key topics that help clarify the structure and function of values.

Values affect behavior through a hierarchy of cognitions. What is the relationship between stable, enduring characteristics, such as values, and a person's situation-specific behavior? For example, will people who have a high rating of *universalism* behave differently toward wildlife from those with low ratings of *universalism*? The value–attitude–behavior hierarchy (VAB) answers this conceptually. VAB proposes that values affect mid-range attitudes and that mid-range attitudes influence behaviors. Values are "abstractions from which attitudes and behaviors are manufactured" (Homer & Kahle, 1988, p. 638). This suggests, for example, that a person with universalism values would express those values in her attitudes on a significant number of topics (e.g., protecting endangered species, women's rights, anti-war, humanitarian causes). These attitudes, in turn, lead to a person behaving in a way that is consistent with such values (e.g., she may donate money, vote, express views in support of these topics). In this regard, we can identify what causes an individual's behavior based on values important to him or her. Moreover, this link between values and behavior has important implications: if we understand a person's values– or, as proposed later in this chapter, her wildlife value orientations – we can anticipate her reactions to a variety of issues.

Attention to the hierarchical nature of psychological concepts has been an important trend of the last two decades. Research by Homer and Kahle (1988), which was highly influential, affected how subsequent research explored VAB relationships. These researchers used structural equation modeling to demonstrate the relationship between values, attitudes toward nutrition, and shopping for natural foods. People with strong *internal* values (including self-fulfillment, fun and excitement in life, sense of accomplishment, and self-respect) are more likely to shop for natural food. People with high *external* values (including sense of belonging, security, and being respected) were less likely to shop for and eat natural foods.

The hierarchical model has been applied in the natural resources field in several contexts. For example, Stern et al. (1995) introduced a hierarchical model that explained environmental concern. They proposed a general-to-specific sequence that included the following:

Position within social structure or institutional structure → Values → General beliefs or worldview or folk ecology theory → Specific beliefs and attitudes → Behavioral commitments or intentions → Environmental behaviors (e.g., boycotting companies that pollute, signing petitions for tough environmental laws).

Fulton, Manfredo, and Lipscomb (1996) proposed a hierarchical model about behavior toward wildlife. These authors proposed that wildlife value orientations affect attitudes that then affect behaviors. Their research detected two orientations: *wildlife protection* and *wildlife utilization*. Tests of their model showed that the protection orientation predicted attitudes and behaviors related to wildlife viewing intentions, while utilitarian value orientations predicted hunting participation.

Other researchers have used the VAB hierarchy models within natural resources and environmental management. In recent years, researchers have used the basic VAB model to examine public attitudes toward ecological restoration (Bright, Barro, & Burtz, 2002), intentions to participate in hunting (Hrubes et al., 2001), contingent valuation of endangered species (Kotchen &

Reiling, 2000), intention to vote for wildland preservation (Vaske & Donnelly, 1999), and recycling and other environmental behaviors among Spaniards (Corraliza & Berenguer, 2000).

Research by Manfredo and Teel (Chapter 8) extended this hierarchy by incorporating it into a macro–micro model of shifting wildlife values. At the individual level, values and value orientations govern people's behaviors. At the macro level, values and value orientations are rooted in the forces of modernization (economic well-being, urbanization, education) that affect lifestyles.

Values form slowly over many experiences. Rohan (2000) argued that values are cognitive structures that organize information about past experiences. This information aids people's evaluations and serves as an analogy for interpreting new information and events; i.e., people interpret new information from prior experiences stored in the cognitive structure.

Psychologists use a concept called *schema* to describe the way people form values and store and retrieve memories (Rohan, 2000; Smith, 1998). Schema are abstract, generalized knowledge derived from a person's experiences. Markus and Zajonc (1985) suggested that schema are subjective theories about how the social world operates. Smith (1998) contended that the development of a schema may be derived from specific processes of the neocortical systems, which occur very slowly through the derivation of observed regularities. Schema form in semantic memory, instead of episodic memory. Values form through the consolidation of many past experiences, and because of this, they are stable. Single events have little impact on them, and once a person forms values, those values are unlikely to change without massive and convincing evidence that conflicts with pre-existing positions.

When schema knowledge is activated, it is entirely engaged and brought to memory. When schema are retrieved, the generalized abstraction formed by the schema is referenced, and traces of specific experiences that built the schema are remembered. For example, injuring a deer in an auto accident may flood a person's mind with a general sense about appropriate treatment of other life and traces of similar experiences that illustrate such principles.

Schema are important because they guide interpretation. They help us interpret events and new information. They direct our attention to certain cues and fill in the gaps of our knowledge when situations are ambiguous, and they guide the retrieval of schema-consistent information. To illustrate, many people in the United States probably have schemas about recreational sport hunting. For these people, the sight of a person dressed in clothes with blaze-orange coloration pumping gas into his vehicle at a local convenience store activates recreational sport hunting schema. Although the observer may have no more information than what she saw, she may infer that: soon many hunters will be in the town and at restaurants, hunters in town are unpredictable and unsavory people, hunters will kill many animals, many animals will suffer, and it will be unsafe to be in the woods. The observer may form a negative impression of the person pumping gas and avoid the gas station. For other individuals, seeing the hunter in town may evoke thoughts that the deer hunting season is approaching, that this provides hunters a time to socialize with friends, and that this provides an opportunity to enjoy the outdoors and shoot a trophy deer and enjoy family traditions. This person may identify with the person pumping gas and talk with him after pulling into the station.

Several researchers apply a hierarchical, schema-based approach to understand human behavior (e.g., Bagozzi, Bergami, & Leone, 2003). D'Andrade (1992) stated "not all schemas function as goals, but all goals are schemas" (p. 31). He suggested that there are three levels of goals that act together to form a person's interpretative system. A person's most general goals are considered *master motives*. Master motives are goals like love, security, and belonging. They are ends that have no more "ultimate goals in sight" (p. 30). Farther down the hierarchy are middle-level schemas, which include things like *my marriage*, *my job*, and *my leisure pursuits*. At the lowest level are schemas about simple objects such as windows, bicycles, and chairs. These do not instigate action unless paired with other schemas.

D'Andrade indicated that schemas need schemas higher in the schema hierarchy to instigate action. For example, a schema about hiking may contain cognitions about how to dress, how to locate trails, how to use a compass, and how to read a map. To activate this schema, it needs to be paired with a broader schema, perhaps one about daily exercise. This broader schema may include beliefs about feeling better, weight control, healthy condition, and being stimulated when immersed in natural environments. In turn, this schema may be related to more general goals about living a happy and longer life.

In sum, schema help us conceptualize the structure of values, the links among values and other cognitions, the means by which people develop values, and how those values function within a person's cognitive hierarchy. Schema also explain why values are stable: they (schemas) help us understand the world through accumulated experience, they help us predict the behavior of other people with similar values, and they form and change slowly.

Values are important elements of cultural transmission. Humans have a unique characteristic relative to other life: we transmit and accumulate knowledge across generations. To maintain traditions, customs, values, and what we generally refer to as culture, we depend on this transmission process.

Value formation illustrates how this transmission process occurs. Values form through repeated exposure to situations (e.g., being told what is right and wrong, hearing stories about right and wrong, seeing or receiving praise or punishment for what is right or wrong, seeing others rewarded or punished, etc.). The formation of values is determined not by just one person or one event, but a myriad of events. Prevalence of thought and customs dictates the formation of values. Massive change is required in many areas of life to change values; this is why values are so enduring. The stability of values allows us to predict human behavior. Predictability is critical for social interaction, cooperation, and sustainability of cultural groups.

Values are linked to prevailing human needs. The concept of needs is fundamental in psychology and has an important relationship to the concept of values. While both are theorized to influence human behavior, they are quite different. Needs have a biological, inherited basis, and, if left unfulfilled, have unfortunate consequences. For example, if the need for safety and security is not met, effective human functioning is threatened. In contrast to needs, values are learned goals and cultural constructions. Hitlin and Piliavin (2004) drew the link between these two concepts by suggesting that values are "socially acceptable, culturally defined ways of articulating needs" (p. 361). Reproduction needs, for example, might manifest themselves through cultural values regarding romantic love and marriage. Schwartz (2004) embraced that view and contended that values are grounded in three basic universals required for human existence: biologically based needs, needs for social interaction, and survival and welfare needs of groups. Because humans require cooperation in fulfillment of those needs, values become critical because they serve as a way to communicate with others and enlist their assistance.

Ron Inglehart, a political scientist, theorized that a worldwide value shift is occurring. He suggested this shift is caused by shifting need states (Inglehart, 1997). Inglehart adopted Maslow's hierarchy of needs as a basis for his explanation. In Maslow's approach, needs are prioritized in this order:

1. Physiological needs
2. Safety needs
3. Belongingness needs
4. Love needs
5. Self-actualization needs

Physiological needs have highest priority, and self-actualization needs have lowest priority.

Maslow (1954) proposed that, as the more basic (higher priority) needs are met, the next higher priority needs become salient. Hence, as physiological needs are met, safety needs become salient and direct a person's behavior. Inglehart applied this explanation to societies. He suggested that with economic growth in post-industrial societies, "post-materialistic" (e.g., belonging) needs replaced existence needs as primary motivators of human behavior. Because need states changed, value structures also changed; i.e., they transitioned from materialist to self-expressive values. Cross-cultural research by Schwartz and Saige (2000) reinforced these findings by showing that as modernization increased – measured through GNP, non-agricultural employment, number of telephones per 1,000 people, percentage of age-relevant people enrolled in secondary education – value priorities changed. As socioeconomic development increased, people put greater emphasis on self-direction, stimulation, benevolence, and hedonism (values similar to post-materialism) and less importance on power, conformity, and security values (values related to materialism).

Research by Manfredo and Teel on wildlife value orientations (see Chapter 8; Teel et al., 2005) examined this explanation as it applies to shifting wildlife value orientations. Their work expanded upon Inglehart's explanation by showing the following:

(1) A statistical association between Inglehart's value measures and measures of wildlife value orientations. This suggests wildlife value orientations fit into the broader context of life values and are affected by similar forces of change.
(2) That the composition of values within a state is strongly associated with modernization variables (urbanization, education, income).
(3) That the Western United States is transitioning from domination wildlife value orientations to mutualism wildlife value orientations. This value shift from subsistence oriented to self-expressive is associated with modernization.

In conclusion, research suggests that values and wildlife value orientations correspond to the prevalent needs within a cultural group. Human needs do not shape all facets of wildlife-related customs, but at a macro scale needs shape societal thought, and societal thought directs behavior.

Clarifying the Relationship Among Concepts of Values, Ideology, and Value Orientations

Ideology has many competing definitions. As used here, ideology is a concept that subsumes groups of attitudes and values (Maio, Olson, Bernard, & Luke, 2003). De St. Aubin (1996) suggested that personal ideology, which he equated with the term *worldview*, is a "large amorphous component of personality that includes...elements such as political orientation, religiosity, value systems, morality...and assumptions concerning human nature" (p. 152). Pratto (1999) had a similar view but emphasized the social nature of ideologies by describing them as consensually held beliefs that enable the people who share them to understand meaning, to know who they are, and to relate to one another. She suggested that social ideologies were reflected in social stereotypes, principles of resource allocation, role prescriptions, origin myths, citizenship rules, and other stories or ideas that define groups. Ideologies, she contended, tend to perpetuate across generations. They maintain power differentials within society; they structure social, legal, and economic practices, and they produce social relationships that reinforce the ideology and culture.

Ideology has many dimensions. *Subjugation-domination* reflects the extent to which people feel at the mercy of, or have mastery over, their surroundings (Kluckholn, 1951; Milton, 1996). *Individualism* versus *collectivism* reflects high value on the achievement of the independent versus the importance of sustaining group cohesion (Triandis, 1995). *Traditional/religious* versus *secular/rational* during industrial periods reflects differences in how we understand and explain human purpose and humans' relationship to the world (Inglehart & Welzel, 2005). These concepts can be used to contrast cultural groups or large segments of society. Douglas and Wildavsky (1982), for example, proposed that hierarchical societies with an ideology of individualism (such as a large percentage of people in the United States) tend to see the environment as highly resilient to human impacts.

How does ideology relate to values? These two concepts are closely related. Rohan (2000), for example, suggested that ideologies can be detected in social value systems. More recently, Schwartz (2006) reiterated an idea espoused by Kluckholn (1951). Kluckholn proposed that *value orientations* reflect the social ideology of a cultural group. Value orientations are discussed in the next section.

Value Orientations

The values literature uses the concept of orientations somewhat loosely. The term value orientations is often used synonymously with value priorities. This use is inconsistent with the original concept of value orientations introduced by Kluckholn (1951). Kluckholn proposed that value orientations represent, at both the individual and group levels, *unity thema* or ethos that capture the personality of a cultural group.

He defined value orientation as "...*a generalized and organized conception, influencing behavior, of nature, of man's place in it, of man's relation to man, and of the desirable and non-desirable as they may relate to man–environment and inter-human relations*" (italics in original; Kluckholn, 1951, p. 411). His research contrasted the value orientations of Mormons (mastery over nature orientations), Spanish-Americans (subjugation to nature orientations), and Navaho (harmony with nature orientations).

Schwartz (2006) revitalized the broad cross-cultural nature of the value orientations concept by proposing three bipolar dimensions that capture *cultural ideals*. He proposed that *embeddedness versus autonomy* involves thought regarding relationships between the person and group; *hierarchy versus egalitarianism* addresses the ways in which people act to preserve the social fabric; and *harmony versus mastery* addresses issues of how to manage relationships with the social and natural world. His findings suggest, for example, that *egalitarianism* and *harmony* are important orientations in Switzerland, Spain, Italy, and Slovenia, while *hierarchy* and *mastery* are high in Israel, the United States, Thailand, and South Korea. *Embeddedness* is high in Nigeria, Senegal, Egypt, and Cameroon, while *intellectual autonomy* is high in France, Netherlands, and New Zealand.

His results also show how orientations affect attitudes. When *embeddedness* is high, the country's prevailing attitudes were to oppose immigrants; countries with embeddedness orientations also disagreed with political activism and membership in voluntary organizations. As would be expected, the opposite trend occurred in countries with high *intellectual autonomy*.

These orientations may affect how people perceive wildlife. Ingold (1994) proposed that between the hunter and gatherer stage of development and the agricultural stage of development, there was an ideological shift from egalitarian to hierarchical/mastery that profoundly affected human–wildlife *and* human–human relationships. For example, in hunter-gatherer societies, all people were perceived as equals, and animals were perceived equal to humans; however, as agricultural societies formed, an elite emerged that controlled other's activities (just as humans assumed control and responsibility over other life forms). Interestingly, Wildavsky (1991) noted a twentieth-century trend in North America toward a more egalitarian society and, in support of this explanation, noted the increase in animal rights activists who proclaimed that animals should have rights like humans.

Manfredo and Teel applied ideology to explain wildlife value orientations. The next section overviews their approach, and Chapter 8 discusses it in depth.

Wildlife Value Orientations

Manfredo and Teel (Chapter 8) proposed that integrating ideology into the VAB model is critical in understanding the meaning people assign to values and will enhance the use of VAB in intergroup and cross-cultural study. They identify two key value orientations that affect relationships with wildlife in North America: *domination* and *mutualism*. In a study that included 19 of the western United States, they found the following:

- The publics of the western United States vary considerably in mutualism and domination wildlife value orientations (See Fig. 6.2, Maps 1 and 2). People were classified by their scores on the domination and mutualism wildlife value orientations. *Traditionalists* were those above the median on domination and below the median on mutualism. *Mutualists* were those who scored above the mutualism median and below the domination median. *Pluralists* scored above the median on both scales. *Distanced* scored below the median on both domination and mutualism scales. The maps show that states on the West Coast, Hawaii, and the rapidly urbanizing southwest have higher populations of mutualists. Alaska, states in the Midwest, and states in the northern intermountain west are predominated by traditionalists.
- Wildlife value orientations, as proposed by theory, were strongly predictive of attitudes toward fish and wildlife issues. To illustrate, people with a strong domination orientation were far more accepting than mutualists of management techniques that result in direct harm to wildlife. Across the 19 different surveys that were administered in the study, 473 different attitude questions were asked of which 71% (337) were shown to be statistically related (via correlation analysis, $p < .05$) to the domination scale, while the mutualism scale was predictive of attitudinal responses 59% (279) of the time. The highest correlations were on issues dealing with direct harm to wildlife,

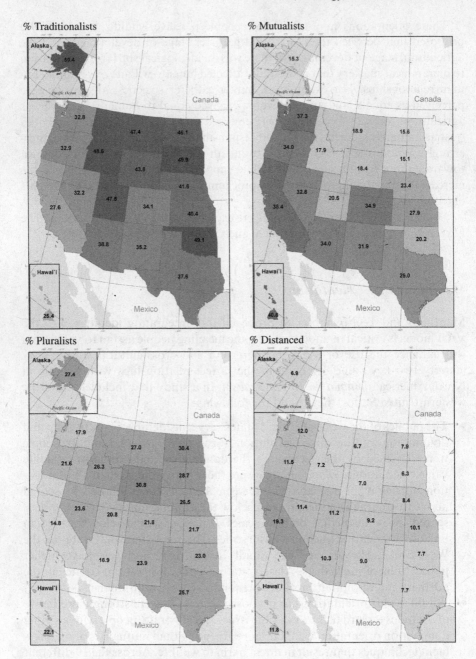

Fig. 6.2 Distribution of wildlife value orientation types across states from a 2004 survey of residents in the western United States

wildlife protection versus human needs and interests, and provision of wildlife viewing or education opportunities.

- Wildlife values in the western United States are shifting from domination to mutualist; this shift appears to be associated with increased modernization (increased education, economic well-being, urbanization). Modernization alone does not change individuals, rather it reflects change that occurs in people's daily lives; this daily change causes intergenerational value shift. In particular, modernization creates a lifestyle where

  - Wildlife are no longer seen as a necessity for survival.
  - People learn about wildlife from indirect sources instead of direct sources.
  - A changing social environment weakens utilitarian views, and people's inherited tendency to anthropomorphize facilitates seeing animals as potential companions in a society increasingly focused on belongingness needs.

Because the value orientations approach integrates ideology, and because ideology is an important concept in describing differences in cultural thought, the wildlife value orientation concept can be used in cross-cultural contexts. To explore this possibility, researchers applied the value orientation concept in different countries (Teel, Manfredo, & Stinchfield, 2007) including China (Zinn & Shen, 2007), Estonia (Raadik & Cottrell, 2007), Mongolia (Kaczensky, 2007), Netherlands (Jacobs, 2007), and Thailand (Tanakanjana & Saranet, 2007).

Using qualitative techniques (Dayer, Stinchfield, & Manfredo, 2007), researchers categorized responses by value orientation concepts. Based on a pretest of the methodology (Dayer et al., 2007), responses were categorized by orientations classified as *materialism* versus *mutualism; concern for safety* versus *attraction, rational/scientific* versus *spiritual/religious*. Additional categories included *Respect* (related to a general value of respect for life) and *Environment-alism* (reflecting the symbolic nature of wildlife in the concern for environmental quality). Conclusions from the effort suggested that commonalities in value orientations could be detected across countries; these orientations allowed for effective categorization of responses and appeared to reflect the meaning associated with comments made. Moreover, the presence of mutualism, strongest in post-industrialized Netherlands, suggests that the association between mutualism WVO and modernization merits examination at a global level.

These findings open other interesting possibilities for future research that would be usefully applied at a global level. Questions these studies may investigate are as follows: What types of orientations are associated with effective conservation policies? What challenges exist in attempting to affect conservation in countries where certain types of orientations are present? Is what we learn in one country with a certain prevalence of orientation types applicable in countries with similar orientation characteristics? These questions merit further exploration.

Conclusion

The concept of values has been used frequently in studies of human–wildlife relationships and will likely be used frequently in the future. Past research has been directed toward trying to develop a parsimonious classification of types of values. Important strides have been made in suggesting the different ways people think about wildlife and how basic thoughts affect observable behavior. Understanding values is a starting point for managing and planning for wildlife (see the Managerial Applications section at the end of this chapter).

Future research should reach beyond these classifications. It would be important, for example, to enhance our understanding of how values and orientations fit within the context of multiple scales. How are values and orientations affected by and how do they affect other aspects of culture (e.g., material, technological, institutional) and the environment? It would also be important to understand how values and orientations form and change within societies. Is it possible to direct the formation of values and orientations? Finally, given their broad natures, the value and values orientation concepts offer an approach for building a broadly generalizable, cross-cultural explanation of human–wildlife relationships.

Summary

- Research on values toward wildlife was among the earliest conducted by human dimensions of wildlife researchers. Among these early efforts, Kellert's values typology is the most enduring approach. It has encouraged additional research, and it affects how managers consider the social dimension of wildlife management. Future research using Kellert's values typology must address both its methodological and conceptual weaknesses.
- Two social psychologists offer predominant theories about values: Rokeach and Schwartz. Rokeach's theory presents a typology of values; values function as desired end states and modes of conduct. Schwartz's theory is replacing Rokeach's. Schwartz developed a typology of opposing value types that inherently is designed to explain differences among people. The environmental and wildlife fields have used Schwartz's approach successfully. In Schwartz's theory, value clusters related to *openness to change* and *self-transcendence* usually spawn more *mutualistic* views toward wildlife when compared to *conservation* and *self-enhancement*.
- Characteristics of values suggest they belong to a hierarchy of cognitions and that values direct behavior by influencing attitudes. Values are culturally directed ways of meeting basic human needs; they are formed slowly through learning, are organized as schema, and are transmitted as part of cultural learning.
- The ideology concept is broader than the value concept and encompasses many ideas about human nature. Value orientations may reveal ideology's influence on a group or the cultural personality of a group. Researchers

propose wildlife value orientations that reflect a domination and a mutualism orientation. Research in North America suggests value orientations are shifting from domination to mutulism. Some international research applies the value orientations approach.

Management Implications

Like many findings borne from social science research, knowledge about values is most useful in how it affects our thinking about a natural resource problem. To illustrate, many state fish and wildlife agencies in the United States are attempting to develop programs that will recruit or retain hunters. If, as suggested in wildlife value orientations research, hunting participation is the product of broad-based value shift that is driven by forces of modernization, recruitment programs are unlikely to be successful over the long term. More specifically, the long-term influences on participation trends are not attributable to the lack of places to go or information about how or where to participate. It is because participation no longer fits within the context of modern social life and the needs of people in post-industrial society. Hence, the problem may not be finding ways to recruit new hunters but how to engage new, emerging interests. In this fashion, values information can inform many areas of the decision-making process. A few illustrations are provided below.

Classifying the diversity of stakeholders Traditional classifications of stakeholders – such as hunters, anglers, and viewers – will always remain useful; however, such labels are inadequate for understanding the diverse array of public interests. To illustrate, research from both Kellert (1978) and Teel et al. (2005) suggests that all hunters cannot adequately be represented in one category. The Teel et al. study suggested that many hunters had strongly held mutualist and domination orientations, i.e., they were pluralists. The interests of pluralist hunters would be different than the more extreme domination-oriented traditionalists. By knowing the variety of value types within the population, the nature of their attitudinal leanings and even the region of their residence, managers are in a better position to represent stakeholders in their decision and be more effective in communicating with them.

Guiding visioning and planning The process of planning is well served by an understanding of the values that are important to stakeholders. For most planning models, there is a direct "plug in" for the values terminology. That is, in most cases, the starting point for planning is to identify the goals that are important to a group. Goals are defined as an expression of social values. For example, a goal for embracing the mutualist orientation might be worded in the following way: To recognize the close connection between life among humans and wildlife, and to engage in activities that promote sustainability of human–wildlife co-existence.

In addition, planning tries to anticipate the future. If, as is suggested in prior research, value orientations toward wildlife are expected to change (as described in the example above), then it is important to develop plans that prepare for this change.

Understanding the basis for conflict and assist in consensus building Theoretical explanations of conflict suggest that the root of disagreement is in differing values among opponents. The goal interference hypothesis suggests that conflict arises when the actions of one individual or group blocks the attainment of goals (e.g., values) for another group or individual (Jacob & Schreyer, 1980). Conflict resolution may depend on finding areas of goal similarity among stakeholders and building these areas to seek compromise.

Understanding attitude strength Attitudinal links to values help us understand the vulnerability of an attitude. In some cases, attitudes represent the values that are important to a person (value expressive). When an attitude is rooted in values, one concludes that this attitude would be very difficult to change. To illustrate, a study was conducted to understand the basis for Coloradoans, votes on a ballot initiative to ban recreational trapping in the state (Manfredo, Fulton, & Pierce, 1997). Findings suggested that the foundation of votes to ban trapping was rooted in wildlife value orientation beliefs that defined what is humane; This suggests it would be very unlikely that attitudes toward trapping could be easily changed.

Understanding fish and wildlife professions Another use of the values concept has been to understand changes in the wildlife profession and professional organizations. Research by Gigliotti and Harmoning (2004), which used wildlife value orientations scales, revealed that professionals in the South Dakota Game, Fish and Parks Department, had, as a group, different WVOs than the general public. These professionals believed that the public had values similar to their own. In other research, a survey of alumni from Colorado State University's College of Natural Resources showed that recent graduates were far more likely to embrace a wildlife rights orientation than would older generations of graduates. These findings lead to important questions like: how well do agencies represent the broad array of stakeholders they serve? How can an agency recruit and retain employees that value diversity? How do employees affect the structure and personality of an agency so that it can serve society effectively.

References

Allport, G. W. (1961). *Pattern and growth in personality*. New York: Holt, Rinehart, & Wilson.
Bagozzi, R. P., Bergami, M., & Leone, L. (2003, October). Hierarchical representation of motives in goal setting. *Journal of Applied Psychology, 88*(5), 915–43.
Bardi, A., & Schwartz, S. H. (2003). Values and behavior: Strength and structure of relations. *Personality and Social Psychology Bulletin, 29*(10), 1207–1220.

Bright, A. D., Barro, S. C., & Burtz, R. T. (2002). Public attitudes toward ecological restoration in the Chicago metropolitan region. *Society and Natural Resources, 15*, 763–785.

Bryan, H. (1980). Sociological and psychological approaches to assessing and categorizing wildlife values. In W. W. Shaw, & E. H. Zube (Eds.), *Wildlife values* (pp. 70–76). Center for Assessment of Noncommodity Natural Resource Vales, Insti. Series Report 1.

Campbell, D. T. (1963). From description to experimentation: Interpreting trends as quasi-experiments. In C. W. Harris (Ed.), *Problems in measuring change*. Madison: University of Wisconsin Press.

Clark, T. W., & Kellert, S. R. (1988). Toward a policy paradigm of the wildlife sciences. *Renewable Resources Journal, 6*(1), 7–16.

Corraliza, J. A., & Berenguer, J. (2000). Environmental values, beliefs and actions: A situational approach. *Environment and Behavior, 32*, 832–848.

D' Andrade, R. (1992). Human motives and cultural models. In R. D'Andrade, & C. Strauss (Eds.), *Human motives and cultural models* (pp. 23–44). Cambridge: Cambridge University Press.

Dayer, A. A., Stinchfield, H. M., & Manfredo, M. J. (2007). Stories about wildlife: Developing an instrument for identifying wildlife value orientations cross-culturally. *Human Dimensions of Wildlife, 12*(5), 307–315.

De St. Aubin, (Ed.). (1996). Personal ideology: Its emotional foundation and its manifestation in individual value systems, religiosity, political orientation, and assumptions concerning human nature. *Journal of Personality and Social Psychology, 17*, 152–165.

Decker, D. J., & Goeff, G.R. (Eds.). (1987). *Valuing wildlife: economic and social perspectives*. Boulder, CO: Westview Press.

Dietz, T. A., Frisch, S., Kalof, L., Stern, P. C., & Guagnano, G. A. (1995). Values and vegetarianism: An exploratory analysis. *Rural Sociology, 60*(3), 533–542.

Douglas, M., & Wildavsky, A. (1982). Risk and culture: An essay on the selection of technical and environmental dangers. Berkley: University of California Press.

Feather, N. T. (1988). Values, valences and course enrollment: testing the role of personal values within an expectancy value framework. *Journal of Educational Psychology, 80*, 381–391.

Fulton, D. C., Manfredo, M. J., & Lipscomb, J. (1996). Wildlife value orientations: A conceptual and measurement approach. *Human Dimensions of Wildlife, 1*(2), 24–47.

Gigliotti, L., & Harmoning, A. (2004, Spring). Findings abstract. *Human Dimensions of Wildlife, 9*(1), 79–81.

Hautaluoma, J., & Brown, P. J. (1978). Attributes of the deer hunting experience: a cluster-analytic study. *Journal of Leisure Research, 10*, 271–278.

Hendee, J. C. (1974). A multiple-satisfaction approach to game management. *Wildlife Society Bulletin, 2*(3), 104–113.

Hitlin, S., & Piliavin, J. A. (2004). Values: Reviving a dormant concept. *Annual Review of Sociology, 30*, 359–393.

Homer, P. M., & Kahle, L. R. (1988). A structural equation test of the value-attitude-behavior hierarchy. *Journal of Personality and Social Psychology, 54*, 638–646.

Hrubes, D., Ajzen, I., & Daigle, J. (2001). Predicting hunting intentions and behavior: an application of the theory of planned behavior. *Leisure Sciences, 23*, 165–178.

Inglehart, R. (1997). *Modernization and postmodernization*. New Jersey: Princeton University Press.

Inglehart, R., & Welzel, C. (2005). *Modernization, cultural change and democracy: The human development sequence*. New York: Cambridge University Press.

Ingold, T. (1994). From trust to domination: An alternative history of human-animal relations. In A. Manning, & J. Serpell (Eds.), *Animals and human society* (pp. 1–22). New York: Routledge.

Jacob, G. R., & Schreyer, R. (1980). Conflict in outdoor recreation: A theoretical perspective. *Journal of Leisure Research, 12*, 368–380.

Jacobs, M. H. (2007). Wildlife value orientations in the Netherlands. *Human Dimensions of Wildlife, 12*(5), 317–329.

Kaczensky, P. (2007). Wildlife value orientations of rural Mongolians. *Human Dimensions of Wildlife, 12*(5), 317–329.

Kaltenborn, B. P., & Bjerke, T. (2002). The relationship of general life values to attitudes toward large carnivores. *Human Ecology Review, 9*(1), 55–61.

Kellert, S. R. (1976). Perceptions of animals in American society. *Transactions of North American Wildlife and Natural Resources Conference, 41*, 533–546.

Kellert, S. R. (1978). Attitudes and characteristics of hunters and anti-hunters. *Transactions of North American Wildlife and Natural Resources Conference, 43*, 412–423.

Kellert, S. R. (1980). American attitudes toward and knowledge of animals: An update. *International Journal for the Study on Animal Problems, 1*(2), 87–112.

Kellert, S. R. (1983). Affective, evaluative and cognitive perception of animals. In I. Altman, & J. Wohlwill (Eds.), *Behavior and the Natural Environment* (chapter. 7). New York: Plenum Press.

Kellert, S. R. (1984). Assessing wildlife and environmental values in cost-benefit analysis. *Journal of Environmental Management, 18*, 355–363.

Kellert, S. R. (1985). Public perceptions of predators, particularly the wolf and coyote. *Biological Conservation, 31*, 167–189.

Kellert, S. R. (1993). The biological basis for human values of nature. In S. R. Kellert, & E. O. Wilson (Eds.), *The biophilia hypothesis* (pp. 41–69). Washington, DC: Island Press.

Kellert, S. (1996). *The value of life: Biological diversity and human society*. Washington, DC: Island Press.

Kellert, S. R. (2000). Values, ethics, and spiritual and scientific relations to nature. In S. R. Kellert, & T. J. Farnham (Eds.), *The good in nature and humanity* (pp. 49–64). Washington, DC: Island Press.

Kellert, S. R. (2002). Experiencing nature: affective, cognitive, and evaluative development in children. In P. H. Kahn, & S. R. Kellert, (Eds.), *Children and nature: Psychological, sociocultural, and evolutionary investigations* (pp. 117–151). Cambridge: MIT Press.

Kellert, S. R., & Berry, J. K. (1987). Attitudes, knowledge and behaviors toward wildlife as affected by gender. *Wildlife Society Bulletin, 15*(3), 363–371.

Kellert, S. R., & Wilson, E. O. (1993). *The biophilia hypothesis*. Washington, DC: Island Press.

King R. T. (1947). The future of wildlife in forest and land use. *Transactions of the North American Wildlife Conference, 12*, 454–467.

Kluckholn, C. (1951). Values and value-orientation in the theory of action: An exploration in definition and classification. In T. Parsons, & E. Shils (Eds.), *Toward a general theory of action* (pp. 388–433). Cambridge, MA: Harvard University Press.

Kotchen, M. & Reiling, S. D. (2000). Environmental attitudes, motivations, and contingent valuation of non-use values: a case study involving endangered species. *Ecological Economics, 32*, 93–107.

Langeneau, E., Kellert, S. R., & Applegate, J. E. (1984). Values in management. In L. K. Halls (Ed.), *White-tailed deer: ecology and management* (pp. 699–720). Harrisburg, PA: Stackpole Books.

Maio, G. R., Olson, J. M., Bernard, M. M., & Luke, M. A. (2003). Ideologies, values, and behavior. In J. Delamater (Ed.), *Handbook of social psychology* (pp. 283–308). London: Springer.

Manfredo, M. J., Teel, T. L. & Bright, A. D. (2004). Application of the concepts of values and attitudes in human dimensions of natural resources research. In M. J. Manfredo, J. J. Vaske, B. L. Bruyere, D. R. Field, & P. J. Brown (Eds.), *Society and natural resources: A summary of knowledge*. Prepared for the 10th International Symposium on Society and Resource Management (pp. 271–282). Jefferson, Missouri: Modern Litho.

Manfredo, M. J., Zinn, H. C., Sikorowski, L., & Jones, J. (1998). Public acceptance of mountain lion management: A case study of Denver, Colorado, and nearby foothills areas. *Wildlife Society Bulletin, 26*(4), 964–970.

Manfredo, M. J., Fulton, D., & Pierce, C. (1997). Understanding voting behavior on wildlife ballot initiatives: The case of Colorado's Amendment 14. *Human Dimensions of Wildlife,* 2(4), 22–39.

Markus, H., & Zajonic, R. B. (1985). The cognitive perspective in social psychology. In G.Lindzey, & E. Aronson (Eds.), *Handbook of social psychology* (3rd ed., pp. 137–230). New York: Random House.

Maslow, A. H. (1954). *Motivation and personality.* New York: Harper and Row.

Milton, K. (1996). *Environmentalism and cultural theory.* London: Routledge.

Nunnally, J. C., & Bernstein, I. H. (1994). Psychometric theory (3rd ed.). New York: McGraw-Hill.

Potter, D. R., Hendee, J. C., & Clark, R. N. (1973). Hunting satisfaction: Game, guns, or nature. In J. C. Hendee, & C. Schoenfeld (Eds.), *Human dimension in wildlife programs* (pp. 62–71). Washington, DC: Mercury Press.

Pratto, F. (1999). The puzzle of continuing group inequality: Piecing together psychological, social, and cultural forces in social dominance theory. In M. P. Zanna (Ed.), *Advances in experimental social psychology* (Vol. 31, pp. 191–263). New York: Academic Press.

Purdy, K. G., & Decker, D. J. (1989, Winter). Applying wildlife values information in management: The wildlife attitudes and values scale. *Wildlife Society Bulletin, 17*(4), 494–500.

Raadik, J., & Cottrell, S. (2007). Wildlife value orientations: An Estonian case study. *Human Dimensions of Wildlife, 12*(5), 347–357.

Rohan, M. J. (2000). A rose by any other name? The values construct. *Personality and Social Psychology Review, 4*(3), 255–277.

Rokeach, M. (1973). *The nature of human values.* New York: Free Press.

Schwartz, S. H. (1992). Universals in the content and structure of values: theoretical advances and empirical tests in 20 countries. *Advances in Experimental Social Psychology, 25*, 1–65.

Schwartz, S. H. (1996). Value priorities and behavior: Applying a theory of integrated value systems. In C. Seligman, J. M. Olson, & M. P. Zanna (Eds.), *The Psychology of values: The Ontario symposium* (Vol. 8, pp. 1–24). Hillsdale, NJ: Lawerence Erlbaum.

Schwartz, S. (2004). Basic human values: Their content and structure across cultures. In A.Tamayo, & J. Porto (Eds.), *Vialores e Trabalho.* Brazil: Editora Universidade de Brasilis.

Schwartz, S. H. (2006). A theory of cultural value orientations: Explication and applications. *Comparative Sociology, 5*, 136–182.

Schwartz, S. H., & Sagie, G. (2000). Value consensus and importance. A cross-national study. *Journal of Cross-Cultural Psychology, 31*(4), 465–497.

Shaw, W. W., & Zube, E. H. (Eds.). (1980). *Wildlife values* (Report 1). Tucson: Center for Assessment of Noncommodity Natural Resource Values, University of Arizona.

Smith, E. T. (1998). Mental representation and memory. In D. T. Gilbert, S. T. Fiske, & G. Lindzey (Eds.), *Handbook of social psychology*, (4th ed., Vol. 1, pp. 391–445). New York: Oxford University Press.

Smith, P. B., & Schwartz, S. H. (1997). Values. In J. W. Berry, M. H. Segall, & C. Kagitcibasi (Eds.), *Handbook of cross-cultural psychology: Social behavior and applications* (2 ed.,Vol. 3, pp. 77–118). Boston: Allyn and Bacon.

Stern, P. C., Dietz, T., & Kalof, L. (1993). Value orientations and environmental concern. *Environment and Behavior, 25*, 322–348.

Stern, P. C., Dietz, T., & Guagnano, G. A. (1995). The new environmental paradigm in social-psychological context. *Environment and Behavior, 27*(6), 723–743.

Stern, P. C., Dietz, T., & Guagnano, G. A. (1998). A brief inventory of values. *Educational and Psychological Measurement, 58*(6), 984–1001.

Stern, P. C., Dietz, T., Kalof, L., & Guagnano, G. A. (1995). Values, beliefs, and proenvironmental action: attitude formation toward emergent attitude objects. *Journal of Applied Social Psychology, 25*(18), 1611–1636.

Stern, P. C., & Dietz, T. (1994). The value basis of environmental concern. *Journal of Social Issues, 50*(3), 65–84.

Sturch, N., S. Schwartz, H., & van der Kloot, W. A. (2002). Meanings of basic values for women and men: A cross-cultural analysis. *Personality and Social Psychology Bulletin, 28*(1), 16–28.

Tanakanjana, N., & Saranet, S. (2007). Wildlife value orientations in Thailand: preliminary findings. *Human Dimensions of Wildlife, 12*(5), 339–345.

Teel, T. L., Dayer, A. A., Manfredo, M. J., & Bright, A. D. (2005). *Regional results from the research project entitled "Wildlife values in the West."* (Project Re. No. 58). Project Report for the Western Association of Fish and Wildlife Agencies. Fort Collins, CO: Colorado State University, Human Dimensions in Natural Resources Unit.

Teel, T. L., Manfredo, M. J. & Stinchfield, H. S. (2007). The need and theoretical basis for exploring wildlife value orientations cross-culturally. *Human Dimensions of Wildlife, 12*(5), 297–307.

Triandis, H. C. (1995). *Individualism and collectivism*. Boulder, CO: Westview.

Vaske, J. J., & Donnelly, M. (1999). A value-attitude-behavior model predicting wildland preservation voting intentions. *Society and Natural Resources, 12*, 523–537.

Vitterso, J., Berke, T., & Kaltenborn, B. P. (1999). Attitudes toward large carnivores among sheep farmers experiencing different degrees of depredation. *Human Dimensions of Wildlife, 4*(1), 20–35.

Wildavsky, A. (1991). *The rise of radical egalitarianism*. Washington, DC: The American University Press.

Williams, R. M. (1968). The concept of values. In D. S. Sills (Ed.), *International encyclopedia of the social sciences* (pp. 283–287). New York: Macmillian Free Press.

Zinn, H. C., & Shen, X. S. (2007). Wildlife value orientations in China. *Human Dimensions of Wildlife, 12*(5), 331–338.

Chapter 7
Cultural Perspectives on Human–Wildlife Relationships

Contents

Introduction

Recently one of my colleagues, an economics professor from Italy, was traveling through the West and stopped to visit. During a dinner conversation, he told me that he and his daughter intended to travel north to Yellowstone National Park (YNP). Yellowstone is generally considered the crown jewel of the American Park System. I told him about the history of our National Park Service, YNP's unique stature in our Park System, and the amazing sights he could expect there. We agreed to have dinner at the conclusion of his trip, on his way back to the Denver airport.

I wanted to know his reaction to the readily accessible views of wildlife in YNP. My trips to Italy's national parks were enjoyable and in my opinion the scenery of the Southern Alps is as awe-inspiring as any place on earth, but for all the beauty there, I found the lack of wildlife striking.

Upon his return, I asked about my friend's experience in YNP and was surprised by his answer. He was amazed at the animals along the roadside. I asked him how people from his home town would react to those sights. "All

M.J. Manfredo, *Who Cares About Wildlife?*,
DOI: 10.1007/978-0-387-77040-6_7, © Springer Science+Business Media, LLC 2008

the lines of cars stopping to watch the animals," he exclaimed. "They created traffic jams that were very irritating. The animals on the road would keep people from going where they wanted to be!"

I reflexively burst out laughing; what a contrast in perspectives! While I would never suggest that this is representative of the Italian view toward wild-life, it does illustrates how risky it is to make assumptions about how animals are regarded in other cultures. What can be said about the different cultural perspectives on human–wildlife relationships?

As noted at the outset of this book, the primary emphasis of this book has been to introduce individual-level concepts that explain people's thoughts and actions regarding wildlife. However, in this chapter, I introduce concepts that have emerged through cross-cultural study that help our understanding of human–wildlife relationships.

Culture is perhaps the broadest and most encompassing concept within the social sciences. As noted by Milton (1996), the traditional view of culture encompasses three realms: action, perception or ideological, and material. Action includes individual's observable behavior. The material realm involves all human-made items and artifacts. Perception or ideology includes the domain of what people think, which includes values, norms, beliefs, knowledge, traditions, customs, and understanding.[1]

While anthropologists debate what should be emphasized when defining culture, they agree that ideology is central (Herzfeld, 2001). In this view, *culture* is the accumulated societal knowledge that is passed between generations. It adapts humans to their social and environmental surroundings. Culture is self-perpetuating and is expressed and reinforced in all areas of life (Salzman, 2001).

The study of culture is traditionally the domain of anthropology, and as noted by Theodossopoulos (2005), "...anthropologists have treated the rela-tionship of people to animals as an analytical tool serving more general theore-tical preoccupations, not as an end in itself" (p. 28). To illustrate the nature of their interests, a noted anthropologist, Radcliffe-Brown, asked in the early 1900s, "Why do the majority of what are called primitive people adopt in their custom and myth a ritual attitude towards animals and other natural species?" (1952, p. 129).

In a more recent review of anthropological literature on human–wildlife relationships, Mullin (1999) suggested that human–wildlife relationships have become a focus of their own, due in part to growing practical concerns. She stated, "With the rise of ecotourism, a global traffic in exotic animals, the spread of factory farming, and transnational conflicts over conservation and the treatment of animals, it is especially important that humans' relationships with animals in one part of the world be considered in relation to those in others" (p. 219).

---

[1] Please note the difference in how the term ideology is used here compared to the way it is used in the chapter on values. Here it encompasses attitudes, norms, affect, and values, whereas in our psychological overview, it is defined and examined as a separable influence.

The purpose of this chapter is twofold. First, it overviews key points in the search for cultural regularities in understanding human–wildlife relationships. In the second part, it overviews theories of cultural change and emphasizes how the cognitive realm of culture (i.e., the knowledge, feelings, and values that individuals possess) fits within the broader components of cultural change.

Different Cultural Perspectives Regarding Wildlife

An understanding of cross-cultural differences in human–wildlife relationships has important practical ramifications. First, many wildlife conservation efforts involve multinational collaboration. Salient examples include many migratory species such as waterfowl, salmon, and whales. Effective collaboration in these cases will be facilitated through mutual understanding of how a society is tied to such wildlife. Second, there is a growing tendency of groups from developed nations (particularly non-governmental agencies) to insert themselves in the conservation issues of developing nations. Effective action in these cases will not be achieved without an understanding of human–animal relationships in the developing nation (as it stands in contrast to the developed nation view) and the socio-cultural conditions that surround human–wildlife interactions. Finally, many wildlife professionals are now beset by a mixture of different cultural perspectives. In North America, for example, wildlife managers must find ways to embrace the interests of indigenous peoples, the growing populations of non-white people, and recent immigrants from countries with different norms about wildlife uses. Inevitably, much of wildlife management will have a cross-cultural element in the future.

Most people would be aware that humans have widely different conceptualizations of wildlife based on their cultural perspective. To the Wasanipi Cree Indian hunters in sub-artic Quebec, animals pursued by hunters were gifts that are like people, given to them by their god (Burch & Ellanna, 1994). To the Karam of the New Guinea Highlands, cassowaries, large ostrich-, or emu-like birds, were considered one's sisters or cross-cousins (Bulmer, 1967). To the Mount Gambier tribe of aborigines in Australia, all objects in the world were organized into clans, such as the crow, pelican, or black cockatoo. To members of the clan, these totemic emblems were sacred beings, and all those within the clan are considered the kin (Durkheim, 1912/1964).

Differences in views of the same animal can vary greatly across countries. For example, in the countries of Sri Lanka, Burma, Thailand, and Kapuchea, there is a strong religious association with elephants, yet in China, there is no religious connection and elephants are exterminated as vermin (Sukumar, 1989). Even among post-industrialized nations, different values toward wildlife are apparent. For example, Kellert (1993) contended that of the leading post-industrialized societies, the Japanese have an interest in wildlife that is "... confined to particular species...admired in a context emphasizing control, manipulation, or contrivance" (p. 66). Conversely, Germans have very idealistic and romanticized attitudes toward animals and the environment, while Americans have a pragmatic view of wildlife. As we consider these differences, we might ask what regularities, if any, might be taken from the diverse array of human–wildlife relationships we observe?

Differences in Human–Wildlife Relationships by Stage of Cultural Development or Structure of Society

Do differences among cultures display a regular pattern of human–wildlife relationships? Stage of culture development is one way that cross-cultural differences toward wildlife have been summarized. For example, Schwabe (1994) proposed that at the earliest stage, which he labeled *folk*, human–animal roles are *fused* (i.e., animals are completely integrated culturally and economically within the social fabric). This occurs in economically undeveloped, pastoral societies with only slight division of labor. In the next stage, *agrarian*, animal roles in society are *prismatic*. In this stage, human society has a greater separation from animals, and animals fulfill multiple utilitarian purposes central to the family and wider economy. Economically developing, village-based societies with plant–animal agriculture show prismatic human–animal roles. These societies have considerable division of labor and differentiation of cultural institutions and social structure. At the most recent stage, which Schwabe refers to as *industrial*, animal roles are *diffracted*; certain species and individual animals have highly specialized roles as direct food providers, close personal companions, and providers of aesthetic and recreational pleasure. Diffracted roles occur in economically developed countries with industry and increasingly intensive agriculture. Industrial societies have extreme division of labor, differentiation of institutions, and compartmentalization of values and loyalties at different levels of social structure.

As a general historical description, Schwabe's categorization has intuitive appeal. But as a model used for understanding human–wildlife relationships, it is superficial. Its stages of development represent a Westernized model of cultural advancement, and it does not help us understand the diverse ways that wildlife attain meaning in human society (i.e., there are many differences in human–wildlife relationships across cultures at a given stage).

A consistent theme in anthropological investigation suggests that the form of social organization provides a foundation for exploring similarities and differences. In particular, Douglas and Wildavsky (1982) proposed a theoretical approach that suggests human engagement with nature is based on the form of social organization within a culture. Organizational form, they proposed, affects how knowledge develops within a society. Social organization can be described using two variables: *grid* and *group*. Grid is high when people's actions are strictly controlled and low when freedom of choice is high. Group is high when people have a commitment to communal interests and low when people act in their own interest.

Cultural perspectives arise through the two-by-two cross-tabulation of these variables. For example, low-grid and low-group societies produce a market form of organization. From this form of organization, a worldview about nature emerges. Those organized in high-grid/high-group social structures are described as *heirarchists*. For them, nature exists to be dominated, and while it is generally robust to human influences, it is perceived to have limits. Low-grid/ high-group people are *sectarians*; they believe nature is fragile. Low-grid/low-group are *entrepreneurs* who see nature as highly robust to human influences. High-grid/low-group people are *fatalists*; they see nature as capricious.

Douglas and Wildavsky applied this model to examine the growth of environmentalism in America. They contended that in the 1960s and 1970s, the two dominant cultural perspectives in America were hierarchical and entre-preneurial. A fringe sectarian group existed. Concern for impacts to wildlife and the environment increased as the sectarian perspective grew. Sectarians, Douglas and Wildavsky contended, endure by opposing the prevailing perspec-tive, and environmentalism provided a unifying mechanism that galvanized opposition to the prevailing views. The entrepreneurs believe that the environ-ment recovers no matter what, and the hierarchists believe that the environment has limits; therefore, caution and central control are needed. The sectarian perspective opposes both hierarchical and entrepreneurial views by suggesting that additional burden on the environment will propel it into a precipitous decline. Fear of pollution and damage to the environment kept voluntary participation within the sectarian perspective high.

Why did sectarian forms of organization arise? The theory suggests that in post-World War II society, access to education and labor-saving devices increased. A large number of educated people could not obtain employment in industry; this led to the expansion of the service sector, which is less tolerant of hierarchical control and pursuit of individual gain.

This theoretical model proposed that the growth of a protectionistic approach toward wildlife grew from a need to sustain the growing sectarian social structure. In fact, Wildavsky proposed that the rise of the animal rights movement is evidence for the growth of an egalitarian society in North America during the mid-1900s (Wildavsky, 1991).

Two final points should be made on the topic of human–wildlife relation-ships by cultural categories. First, anthropologists have an ongoing debate about the most effective way to classify cultures. Bird-David (1990, 1992) and Milton (1996) found inadequacies with current cultural classification systems and suggested that views toward the environment and wildlife should actually play a more central role in the creation of cultural classifications. They suggested a classification of human's perception of environments (including wildlife) as an appropriate replacement for the traditional style of classifying cultures as hunter-gatherer, pastoralists, agriculturalists, and industrialists. The existing system could be replaced with a classification of *human ecologies*, which is based on the way that environments are understood by their participants (e.g., passive environments, protective environments, fragile

environments, vindictive environments). For example, many pre-industrialized societies believe they live in a powerful environment; this environment
provides for them, and they must reciprocate or face consequences. In contrast, a prevalent view in post-industrial society is that people live in a fragile
environment that needs their protection.

Finally, it should be noted that there is a tradition in anthropology that
suggests that explanations of culture must arise from within that culture. At an
extreme, this is the position of the post-modern tradition of anthropology. Post-
modernists propose that scientific explanation of culture is not feasible. They
believe that any imposition of another's language or mental structure in culture
description interferes with findings obtained. They believe there are many
realities (based on one's perspective) and no one version should predominate.
This view suggests broad-based classifications are impossible; each situation
and each interpretation of that situation is unique.

Differences by Religious Orientation

Differences in human–wildlife relationships among contemporary societies
have sometimes been cast as a reflection of their religious orientation.
White (1967), for example, traced problems of a growing ecological crisis in
the United States to the domination perspective of Judeo-Christian religions,
which contrasts with religions of Eastern cultures. Instead of the hierarchical
approaches of western religions, Hinduism sees humans as part of nature. Dwitvedi (2001/1992) described Hinduism as placing a premium on life's sanctity; God
has dominion over all life, and humans have no dominion over their own lives or
non-human life. Hindus believe that the Supreme Being was reincarnated in the
form of various species. In Hinduism, to not eat meat is considered both appropriate conduct and a duty. Buddhism offers a somewhat similar view. According
to De Silva (2001/1987), Buddhism is devoted to a way of life that eradicates
human suffering. The code of ethics for Buddhists requires avoidance of injury to
all. It prescribes the practice of loving kindness toward all creatures. The notion
of karma and rebirth "prepares the Buddhist to adopt a sympathetic attitude
toward animals" (p. 258). Finally, in the Islamic religion, Deen (2001/1990)
suggested, other living creatures are worthy of protection and kind treatment.
Hindu and Buddhist religions are seen as ideals for those who oppose the western
view of the environment (Callicott & Ames, 1989; Knight, 2004).

Irrespective of the specific religion considered, the transition from tradi-
tional/religious to scientific, rational-based reasoning (Weber, 1948) is thought
to have had an important impact on how humans view nature and on human–
wildlife relationships. Gellner (1988) proposed two conflicting forms of cultural
knowledge: social knowledge and referential knowledge. Social knowledge,
prevalent in hunter-gatherer societies, creates group cohesion. This type of
knowledge is norm-focused and strongly associated with ritual. For example,

one might believe that an illness was caused by a person breaking a tribal norm, such as sharing food. The explanation for this malady reinforces behavior that increases or maintains group cohesion. Referential knowledge is the second form of knowledge. It seeks objective knowledge that explains cause. According to Gellner (1988), the rise of referential thinking enormously impacted societal development. Post-industrial society arose from the realization that systematic investigation of nature increased output. This created culture "...which no longer accepts its own concepts as ordained from on high, but which chooses its own, and endows them with only a conditional authority (p. 126)."

Milton's (1996) account of an incident in the Kasigau culture – located in Kenya, Africa – contrasts religious and scientific explanations of human–wildlife interactions. The quote was obtained at a time when the Kasigau culture was in transition away from traditional views with the rise of Christianity and the influence of British scientific education:

> A well-educated man, a devout member of the Anglican Church, was walking along a path...when he was confronted by a cheetah. The animal did not attack, but in his surprise the man stumbled and cut his leg. No one could remember a cheetah having appeared in the village before. The "traditionalists" assumed that some sorcery was at work, that the animal had been sent to cause harm. The non-Anglican born-again Christians prayed for the evil influence of Satan to leave their community and assumed that the man's own faith in God had saved him from greater injury. The man himself thought it likely that bush fires on the plain below had driven the animal out of its normal range. He considered it nonsense to suggest that it had evil intent because it was, after all, just an animal. (1996, p. 122)

Contrasting views cause people to avoid or explain hazardous events differently. The traditional view encourages people to seek reasons for adverse events within society (who is the sorcerer, why did that sorcerer act?), while the second view explains the event as resulting from the cheetah's own needs created by the fires.

A Trend Toward Human–Nature Separation Affects Human Views of Wildlife

How humans classify objects reflects their ideology and their society. Anthropology has suggested that what is considered human versus non-human animal, nature versus cultural, and wild versus domesticated (1) varies among cultural groups, (2) suggests a great deal about the group's relationships with the natural world, and (3) suggests something about the culture's social structure and social relationships.

The ideological separation of humans and nature has been cited as a particularly critical occurrence in shaping human–wildlife relationships. This separation occurs as cultures become more complex in their structure and organization (Trigger, 1998). Cultures organized at a tribal or clan level have little conceptual separation of the person from nature. Osborne (1990), for example, studied human relationships with animals among the U'wa, who reside in the eastern Andean area of Venezuela and Columbia. Her analysis of myths reveals that the U'wa have a cosmology in which they do not rigidly divide themselves from nature. All living things are perceived mortal, but birds, bees, reptiles, and aquatic species are near-immortal; to survive, mortals must continually remind the near-immortals to replenish the universe.

According to Ingold (1994), the human–nature separation fueled cultural expansion. In hunter-gatherer societies, humans and animals are seen as "fellow inhabitants of the same world, engaging with one another not in mind or body alone but as undivided centers of intention and action, as whole beings" (p. 18). Within hunter-gatherer societies, humankind's role is to serve nature. In contemporary western views, humans control nature and are responsible for the survival or extinction of wildlife species. With domestication, Ingold describes a transition from trust to domination in the human–animal relationship. In hunter-gatherer societies, humans perceive themselves in a relationship of mutual responsibilities. Animals present themselves to the hunter, and hunters will not be abusive or wasteful of the wildlife (or else wildlife will no longer present itself). Harvest is "a moment in the unfolding of a continuing – even lifelong – relationship between hunter and the animal." Conversely, the pastoralist, while caring for the animal, assumes no reciprocal relationship with animals and is in complete control of the animal as its protector, guardian, and executioner. Ingold suggested that the transition from trust to domination that pervades human–wildlife relations also marked a transition of human–human relations. The notion that our relationships with wildlife are a mirror for human relationships is a recurring theme in the anthropological literature (Mullin, 1999).

Willis (1990) noted that recent trends in post-industrial society toward environmentalism and ecology represent a migration back toward a view that reduces the separation of humans and nature. In Willis' view, Western cultures are entering a *neototemic phase*. Totemism, which is explained in the next section, has received significant attention from anthropologists.

Totemism. Totemism is one of the earliest and most enduing topics of study in anthropology. Mithen (1996) claimed that the study of totemism formed the core of social anthropology during the nineteenth century, and Willis (1990) suggested that the concept was critical in helping anthropology emerge from a level of phenomenological description toward a genuinely scientific discipline. People within a totemic society believe they have descended from, and are part of, the same breed as a particular species of plant or animal. People within a totemic clan are bound by a common heritage, by an obligation to each other, and by a common faith in the totem. In the totemic animal, a person sees both his brother and his god-like ancestor. The totem clarified in-group and out-group classifications, defined who is friend and enemy, and what is good and bad. The totem represented both a social system (people were organized into groups through the clan membership) and a religious system (the totem was involved in rituals and worship)

(Willis, 1990). Mithen (1996) contended that totemism, broadly defined, was a universal among human hunter-gatherer groups and was pervasive in human society since the Upper Paleolithic Period (about 40,000 YBP). There is little evidence of totemism in earlier periods of the Paleolithic, providing partial evidence as to the time that sophisticated, integrated human thought processes emerged.

Among the many explanations regarding the observed regularity of totemism, Lévi-Strauss's theory attained the most attention (Willis, 1990). Totemism, he proposed, is the universal human tendency to use analogical reasoning. Observations of natural species provide a basis for considering relationships among human groups. The differences among wildlife and their habits provide humans conceptual support for their own social differentiation and social classification. Hence, Lévi-Strauss contended that animals are not just "goods to eat," they are "goods for thinking"; they provide accessible theories of relations between human groups (Leach, 1970; Lévi-Strauss, 1963).

Lévi-Strauss proposed that the widespread presence of totemism reveals something far more basic than similar social practices. It reveals the pan-human cognitive process that creates systems of meaning using oppositional differences. It is evident in the classifications such as life-death, light-dark, and human–non-human. Social phenomenon, Lévi-Strauss proposed, can be examined as a system of signs, which are organized as semantic units. Lévi-Strauss's explanation had a significant effect on the ongoing examination of totemism and the symbolic role of wildlife in culture (Willis, 1990).

Symbolic versus material explanations of human–wildlife relationships. In a 1985 review of anthropological studies about human–animal relationships, Shanklin summarized two types of research on human–wildlife relationships. These two approaches emerge from separate traditions in anthropology and focus on different aspects of human–wildlife relationships; they split along the lines of Lévi-Strauss's dual view of wildlife as goods to think and goods to eat. The first approach explores the utilitarian or functional basis of animals in human societies. This approach is associated with the cultural ecology tradition that later emerged as political ecology and ecosystem approaches to depicting culture. The utilitarian approach examines culture as the result of the dynamic process of humans adapting to the environment. Relationships with animals are shaped in a way that facilitates such adaptation. The classic example of a utilitarian approach to examining human–animal relationships is Harris's explanation of Indian Hindu beliefs about cows (Harris, 1974, 1999). With more than 200 million cows in India and significant issues of poverty and starvation, it appears incongruent to have strict prohibition on their slaughter and consumption. Harris, however, explains that the development of this religious practice is appropriate for India's ecology and technology (i.e., a very utilitarian reason). He showed that the taboo exists because protecting cows was more efficient than slaughtering them. That is, the most efficient use of these cows was plowing, fuel and fertilizer (using dung), dairy products, and for food scavenged by the lower caste from dead cows. Harris proposed that the adoption of this belief occurred through a slow process by which those who did not slaughter cattle survived natural disasters and those who used them for meat lost their farming ability. As awareness of this trend spread, an informal taboo emerged about eating beef that was later followed by practices codified by the priesthood (see Simoons, 1979, for criticisms of this explanation).

The second approach anthropology has used to explore human–animal relationships is through human's use as symbols. This approach studies metaphors and prohibitions, taxonomies and cultural classifications of animals and people, and ritual practices such as sacrifice. Leach (1964), for example, suggested there was a universal tendency to make ritual and verbal associations between eating and sexual practices. He found a correspondence between the way animals are categorized by edibility and the way in which humans are categorized with regard to sex relations.

Douglas's work on the Lele of Central Africa and their regard for the lesser scaly anteater (i.e., the pangolin) explored the use of symbolism in human–wildlife relationships (1957, 1990). Among the Lele fertility cult, the pangolin is regarded as one of the most powerful natural spirits and giver of fertility and good hunting. Douglas linked the ritual rites involved in the catch and consumption of the pangolin to the Lele theories of sickness and health (paralleling violations of rules about social relationships). This was further linked to the Lele's conceptual classification of the pangolin as an anomalous animal (i.e., it did not fit easily into their categories of animals).

For Douglas (1990), how people differentiate and classify animals is a source of metaphors for thinking about differences among humans. The Lele have complex rules prohibiting different types of animal meat for different social categories of people. The categories of these meats reinforce distinctions among male and female, child and adult, living and dead, religious initiates and lay people. In her work with the Lele, Douglas (1990, p. 34) concluded,

> Observing the intricate rules about what an individual human can eat or not eat with safety among animals has a strong practical interest. The daily menu, which differentiates categories of humans by their diet sheet, is the surface appearance of deep theory about life and death and health and sickness.

Douglas' explanation for human society mirroring conceptions of the natural world is more practical than Lévi-Strauss's. While Lévi-Strauss proposed that what is observed in the natural world serves as conceptual evidence to support the arrangement of human social structures, Douglas (1990) argued the opposite. She argued that humans use knowledge of their own world to guide their understanding of the relationships in the natural world.

Mullin (1999) noted that research that integrates both the utilitarian approach and the symbolism approach is becoming more prevalent. For example, Pálsson (1990) described the effects of transitioning from subsistence fishing to commercial fishing in Iceland. In pre-Christian society, fish, explained Pálsson, were a pervasive symbol in Icelandic folklore. Myths and folktales emphasized the contrasts between the opposing worlds of land and sea. Categories of sea creatures by land/sea and human/non-human divisions connoted conditions of good/bad catch and safety/danger. People were at the mercy of supernatural forces; they were small pawns manipulated by aquatic beings. Qualities of "fishiness" were transitory among fishermen. Others would rate the fishing foreman on the number of trips he made, his bravery, and his cleverness (because the number of fish caught by each foreman was believed to be beyond his control). When the society changed from a subsistence society to a capitalistic, market-driven society, these old mythologies and metaphors became obsolete. There was a reversal of roles of humans to fish in the cosmic order; i.e., humans were in control of fish. New ideologies emerged as did new social roles. Fishing success was judged by the prowess of the boat captain, which Pálsson described as "the skipper effect."

In addition to integrating symbolic and utilitarian explanations of human–wildlife relationships, Pálsson's work illustrates another important consideration: how symbolic systems change over time. In Pálsson's model, the mode of economy changed ideology. This theme is consistent with models of cultural change described in the next section.

Culture Change and Ideology

Perhaps one of the most important promises of the social sciences is to help deal with the rapid global societal change that is occurring. On one hand, there is a need to anticipate and adapt to this change. For example, the previous chapter suggests that people's changing value orientations affect their leisure time involvement with wildlife. Hunting in North America is declining while wildlife viewing is increasing. The wildlife management institutions in North America, organized primarily to facilitate recreational hunting, must now face the challenges of a changing society.

Affecting change is more challenging than adapting to change. Rapid global modernization affects the sustainability of both natural and human environments. According to the Millennium Ecosystem Assessment (2005), species loss is accelerating at an astounding pace. Global warming is predicted to have catastrophic effects in all life and, without changing the warming trend, virtually all aspects of natural environments and human environments will be significantly altered. Recent assessments say that the ability to affect carbon emissions (and hence warming) is within our technological capabilities (Mendelsohn & Sachs, 2006). Questions remain about how to mobilize social action to create this needed change. Because of this, we must understand how thought processes change within society. Can human thought be affected at a worldwide level to create change? To help us consider this issue, the remainder of this chapter is devoted to culture change theories and emphasizes the role of cultural thought in the change process.

An Historical View of Material Theories of Culture Change

The topic of culture change has been central to anthropology since its inception. It was of particular interest in archaeology, and given the nature of this discipline (using material remains to draw generalizations), it is unsurprising that materialist theories of culture change have become prominent. My brief summary of culture change looks primarily at materialist theories.

Different traditions in anthropology disagree whether broad-level generalizations about culture are possible. For example, the Boas tradition in anthropology emphasized that customs develop in ways unique to specific cultural groups. Although Boas saw environmental and psychological factors as important in shaping culture, the particular historical circumstance of a group was the most important factor in explaining the current situation. Boas stated, "The phenomenon of our science are so individualized, so exposed to outer accident that no set of laws could explain them" (1932, p. 612).

In contrast to Boas, the functionalist tradition of the early 1900s proposed that the ideological portion of culture is not a random phenomenon. Ideology serves an important purpose by attaining group order and directing behavior in a way that meets human needs. The functionalists emphasized culture's role in serving both the psychological needs of individuals (Malinowski, 1939) and in maintaining cultural institutions and social solidarity (Radcliffe-Brown, 1952). The predictive ability of functional explanations was shown to be weak, in part due to the multitude of ways that ideology can be structured to meet needs (Salzman, 2001). That did not diminish the notion that ideology serves a purpose. According to Butzer (1989), the cultural ecology approaches that followed functionalism, described cultural behavior by the functional role it fulfills.

In the mid-1900s, the emergence of cultural ecology emphasized how the natural environment shapes cultures. This tradition models culture change as the result of complex interaction among components of culture and the natural environment. Leslie White (1953), for example, proposed that culture is a function of technology and energy. Culture, he proposed, becomes more complex as the amount of energy per capita increases. The amount of energy available depends largely on the available technology within culture. Over time, more complex cultures outcompete less complex cultures. Julian Steward (1955) also championed an ecological approach to understanding culture change. Steward assigned a critical role to the environmental settings in which cultures emerge. Steward looked for similarities in the core of culture (economic, political, religious aspects) when comparing cultures at a similar developmental level and proposed a multi-linear path of cultural evolution.

The emphasis of all these theories of cultural ecology was on the material forces of culture change with origins in the writings of Karl Marx (Salzman, 2001). Marx explained stages of social evolution based on the modes of production within society. Marx saw social institutions as a function of the mode of production and ideologies (e.g., values) as a function of institutions and the mode of production. For example, hunters and gatherers (mode of production) led to family based kinship groups (social institution). In comparison, chiefdoms were based on a system of redistribution of goods where goods were given to a central figure and redistributed. Social organization shaped by this form of production was hierarchical with distinct lineages.

Marvin Harris (1979, 1999) built upon the prior theories of White and Steward by presenting cultural materialism. As recently as 1998, Trigger noted that cultural materialism has been "theoretically, the most sophisticated and durable version" of theories of culture change (p. 131). Cultural materialism theory proposes three levels of culture: *infrastructure,* which addresses means of cultural production and reproduction; *structure,* which includes political and social institutions; and *superstructure,* which includes ideology, values, and beliefs (Harris, 1999).

Harris proposed a principle known as the *primacy of infrastructure* in culture shift. This suggests that changes in structure and superstructure emanates from changes in infrastructure (Ferguson, 1995). With this approach, Harris considered infrastructural variables (demography, mode of economic production, technology, and interactions with the environment) first when seeking explanations for why other areas of culture have changed. Harris stated,

> The principle of the primacy of infrastructure holds that innovations that arise in the infrastructure sector are likely to be preserved and propagated if they enhance the efficiency of the productive and reproductive processes that sustain health and well-being and that satisfy basic biophysical needs and drives. (1999, p. 142)

The forces of change mirror natural selection processes in genetics. Innovations arise at the individual level and are selected for or against based on how well they facilitate environmental adaptation at the broad systems level.

Murphy and Margolis (1995) illustrated how this approach explains the rise of feminism in postwar America. Despite the inconsistency with prevailing ideology, married women entered the workforce in record numbers in the 1950s. The feminist ideology, which emerged a full decade later, was the result, not the cause, of women's entry into the workforce. High inflation, which increased the demand for labor, was the material reason women entered the workforce.

Critics of the cultural ecology tradition emphasize a number of issues with the approach (Swingewood, 1998; Trigger, 1998). Cultural ecology is seen as overly deterministic; it proceeds on the untenable assumption that negative feedback produces adaptive change that yields homeostasis, security, and stability (i.e., some cultures do not adapt), it asserts indefensibly that all social ideology has a functional basis, and it implies a fatalistic view of social change unaffected and undirected by individual choice.

Recent approaches originating in cultural ecology include both political ecology and ecosystems modeling (Abel & Stepp, 2003; Butzer, 1989; Scoones, 1999). An ecosystem approach advocates a complex-system science; it studies the self-organization, the uptake, flow, and dissipation of energy, and the pulsating, chaotic nature of change and adaptation (Abel & Stepp, 2003). An ecosystems approach (like a political ecology approach) recognizes the important concept of scale, which places emphasis on the temporal, geographic, organizational boundaries of resolution in an investigation. The focus on scale emerges from an underlying assumption of the interconnectedness of phenomenon across such dimensions. Theoretical explanations are encouraged to reach across these scales.

To illustrate an ecosystemic approach, Galvin, Thorthon, Roque de Pinho, Sunderland, and Boone (2006) called for mathematical modeling of the complex relationships among biophysical and human systems. Such models are computer-based "representations of particular facets of reality" and are useful because they provide a way to synthesize what is known across multiple disciplines. These researchers developed a simulation modeling the Ngorongoro Conservation Area in northern Tanzania. Their model simulates the interaction of communities of people, livestock, and wildlife landscape and allows predictions to be made under different scenarios. In the model that Galvin et al. constructed, they used an *agent-based* approach to model community decisions about land use (i.e., community decisions were the result of aggregating the decisions made by households). This model showed that, given the current area allowed for agriculture and population growth levels, the Maasai residents would depend on outside sources for 25% of their needed calories. Doubling the land available for agriculture among poorer households would greatly improve food security and have negligible impacts on wildlife.

Eric Wolf (1972), a student of cultural ecology theorist Julian Steward, introduced the term *political ecology*. Political ecology is now popular because it has a strong, applied agenda. That agenda is obvious in Robbins (2004), who suggested that political ecology attempts to explain links in the dynamics of social and environmental systems by identifying causes of problems, such as starvation, soil erosion, human health crises, biodiversity decline, and other conditions where some people exploit others or environments for limited gain at collective cost. The approach stresses that ecological systems are political and presumes that there are less coercive, more collective, and sustainable ways of doing things. A political ecology approach might begin with a question, such as *why are fish stocks declining*, and pursues an explanation by examining different social and ecological variables at a progressively broader scale. For example, the individual's decision to fish might be examined in the context of the community's social structure and that social structure's power distribution relative to access to resources. Access to resources might then be linked to national policies and people at the national level directly or indirectly targeting fishing. This may then be placed in the context of global markets for fishing and tourism.

Political ecology provides a well-suited paradigm for the conservation profession's growing global perspective. In the global perspective, issues of social equity and resource use and degradation are glaring. Robbins (2004) noted,

> ....the ongoing, small-scale, empirical research projects conducted by countless nongovernmental organizations (NGOs) and advocacy groups around the world, surveying the changing fortunes of local people and the landscapes in which they live, probably comprise the largest share of work in political ecology. (p. 13)

Culture Shift in Post-industrial Society

Much of the impetus for global change emanates from the rise of industrial society and the transition to post-industrial society. In the remaining pages of this chapter, I now examine a materialist view of this cultural shift.

In first introducing the term "post-industrial society," Daniel Bell (1973) suggested that the move away from industrial society was a worldwide phenomenon associated with changes in technology and the growth of a service economy. In this framework, post-industrial society was contrasted with its forerunner *pre-industrial* and *industrial* phases. A pre-industrial sector is primarily extractive; its economy is based on agriculture, mining, fishing, timber, and other resources such as oil and gas. An industrial sector uses energy and machine technology to manufactures goods. The post-industrial sector focuses on information and its processing; it emphasizes telecommunications and computers to channel information and knowledge.

The societal transition (from pre-industrial to post-industrial) that Bell described has different daily experiences, and these different experiences create different world views. In pre-industrialized society, world views are a *game against nature* in which one's sense of the world is conditioned by the natural elements (seasons, storms, drought, soil fertility, floods, etc.). In an industrialized environment, people's daily activities are removed from the natural environment and involve a manufactured world. Bell described the industrialized world view as "a game against fabricated nature." People with this view have a lower dependence on nature and believe they must survive in a technical, mechanical, bureaucratic world. In post-industrialized society, the world view is "a game between persons." Daily life involves information exchange, and given the growing service sector of society, world views are focused around dealing with other people.

Bell (1973) explained that as these modes of life evolve they do not necessarily replace one another: "...the new developments overlie the previous layers, erasing some features and thickening the texture of society as a whole" (1973, p. xvi). In the post-industrialized world, the agrarian and manufacturing sectors are overlaid with an information technology sector, increasing the diversity and complexity of society. The three world views proposed by Bell (i.e., game against nature, game against manufactured world, game against people) provide the foundation for different views of an ideal future regarding use of the environment.

Post-modernization Theory

Ron Inglehart (Inglehart, 1990, 1997; Inglehart & Baker, 2000; Inglehart & Welzel, 2005), a political scientist, presented a contemporary version of Marx's modernization theory that offered explanations for societal value shift in contemporary cultures. His theory proposed:

> ...Socioeconomic development *does* tend to propel various societies in roughly predictable direction. Socioeconomic development starts from technological innovations that increase labor productivity; it then brings occupational specialization, rising educational levels, and rising income levels; it diversifies human interaction, shifting the emphasis from authority relations toward bargaining relations; in the long run this brings cultural changes, such as changing gender roles, changing attitudes toward authority, changing sexual norms, declining fertility rates, broader political participation and more critical and less easily led publics. (Inglehart & Welzel, 2005, p. 19)

Inglehart argued that there was a general pattern of ideological shift from traditional/religious during pre-industrial periods to secular/rational during industrial periods to emphasis on self-expression values in post-industrial periods (Inglehart & Welzel, 2005). Inglehart's (1997) empirical work supports this proposal; it shows that socioeconomic development drives a shift from materialist to post-materialist values in post-industrialized nations. Inglehart argued that people's prevailing need states, influenced by economic advancement, create this cultural shift. Materialist values arise when existence needs prevail during critical times of value formation (e.g., one's youth). As improved economic well-being alleviates those needs, post-materialist values emerge emphasizing concerns related to belongingness and quality of life. Other cross-cultural findings support the basic tenets of Inglehart's argument (Schwartz, 2006; Schwartz & Sagie, 2000).

Inglehart (1990, 1997) adopted the basic principle of modernization that economic, cultural, and political changes occur simultaneously in coherent patterns, changing the world in predictable ways. However, he makes a modification to the earlier notions about modernization, which were generally discredited by theorists critical of Marx. First, Inglehart suggested that change caused by modernization does not produce a linear path for all societies as originally suggested. Instead, several paths are possible. His notion of post-modernism suggests there is a new trajectory that occurs in post-industrialized societies. Change, he proposed, does not necessarily culminate in democracy; other forms of government are possible. Second, modernization was also criticized for its ethnocentric advocacy of westernism. Modernization trends Inglehart observed in Asia suggested to him that the phenomenon is applicable worldwide. Finally, he proposed that modernization is not deterministic in the sense that economy *causes* the modernization process or that ideology *causes* it. Rather, change occurs through reciprocal interaction among these systems of ideology, economy, and political structures.

Inglehart's theory provides useful but, in some cases, deeply contested conclusions. For example, Inglehart (1990) proposed that shifting values result in a decline of elite-directed political mobilization and a rise of elite-challenging issue-oriented groups. This helps explain trends in the natural resource and wildlife decision-making arena which Decker, Brown, and Siemer (2001) described as a transition from authoritarian to citizen co-management modes of decision making. This transition was driven by the growth in elite-challenging, non-governmental activist groups and effective challenges to the traditional decision process– such as new laws, ballot initiatives, and lawsuits.

By contrast, Inglehart also proposed that the rise of environmentalism is largely a post-modern concern for quality of life (unrelated to materialist needs). Brechin and Kempton (1994) challenged this. They noted that a worldwide interest in environmentalism, which cuts across developed and developing nations, is supported in both materialist and post-materialist countries. Abramson (1997) offered rebuttal by showing that, while support for environmentalism is strong across both materialist and post-materialist countries, it is strongest among post-materialists countries. This debate, however, raises appropriate concern about how environmentalism is characterized and why it has arisen.

While recognizing weaknesses in Inglehart's model, it is particularly useful because (a) it is based on quantitative data collected worldwide and (b) it provides useful insights as to how and why ideology is changing in response to material forces in the twenty-first century. As shown in the final chapter of the book, Inglehart's approach suggests ways that we can expand our exploration of human–wildlife relationships at a global scale.

Ideology and Culture Change

What can we summarize from this theoretical overview about culture change and ideology's role (attitudes, values, norms, customs, etc.) in the cultural change process? First, these theories emphasize the adaptive role of ideology. As conditions surrounding a society change, they create stresses, and ideology changes to respond to these conditions. Sometime adaptation is effective, sometimes it is not. Buckley (1967) considered societies as *complex adaptive systems,* and Carniero suggested that the evolution of culture can be viewed as a succession of adaptive changes made by a society as it adjusts to changes in its physical and social surroundings (2003, p. 180). Culture is the cumulative knowledge of a social group; it is adopted and carried forward from generation to generation because of its ability to assist humans in adapting to their environments. The repeated testing and retesting of traditions and beliefs over time proves their adaptive advantage in serving group interests (Richerson & Boyd, 2006).

Second, these theories have the prevailing view that ideology follows changes induced by material causes. One obvious problem with this view is that it might encourage the assumption that the direction of culture is determined entirely by external forces and not human volition. This could encourage "a dangerous myth of helplessness" (Trigger, 1998, p. 150). However, it is unclear whether society is entirely at the mercy of infrastructural factors of change. Even Harris (1999) emphasized the importance of human innovation in the adaptive process. In his view, human creativity provides the innovation that is "tested" in the context of infrastructure. Furthermore, Inglehart suggested that there is only a strong interrelationship, not necessarily a causal direction. As implied by Pálsson (1990), material forces provide the sideboards, both constraints and impetus, but response is also dictated by the social relations of people and traditions. As Trigger states, "...as long as we remain uncertain about precisely how important a role choice plays in human affairs, it is in our interest to behave as if it plays a significant one" (1998, p. 187).

Third, exploration of the topic of cultural change involves a complex phenomenon. New approaches are being explored that attempt to represent cultural change as an intricately interwoven, multifaceted phenomenon. These new approaches embrace a multiple-disciplinary approach and involve explanations that reach across multiple scales. In short, simple explanations will be elusive.

Fourth, there appears to be a predictable ideological trend as societies have moved through industrial to post-industrial phases. This shift, as we shall see in the last chapter, profoundly affects human–wildlife relationships.

Summary

- Human–wildlife relationships vary considerably across cultures. Stage of cultural development, the culture's social organization, and the type of religion associated with the culture have been used to describe these variations.
- How societies create classifications is reflective of their human–nature relationships. Hunter-gathers saw themselves as part of nature. The shift from hunter-gatherer to pastoral society created a perceived separation of humans and nature. This separation fostered a domination worldview where humans attain mastery over other life.
- The phenomenon of totemism is of central interest in exploring the nature of human–wildlife relationships. Totemism, in which humans see themselves of a common lineage with animals, shapes the religious and social system of a society. The widespread totemism of hunter-gatherer societies may indicate the pan-human tendency of analogical reasoning.

- From a cultural perspective, human–wildlife relationships are explored in two ways. One way focuses on the utilitarian connection of society and wildlife. This approach, which is associated with cultural ecology, studies economic and material relationships to cultural development. The other way focuses on how societies symbolize wildlife. Such symbolism appears in myths, metaphors, rituals, etc., and is associated with theories the society holds about the world and social structures.

- Because our world is rapidly changing, we would benefit from a better understanding of what drives cultural shift and the role of societal thought in that process. The cultural ecology tradition and modernization theory provide a foundation for exploring cultural change as a complex phenomenon that links multiple social and biological factors. Cultural ecology and modernization theory view issues in the context of various scales (e.g., individuals within communities, with nations within a global context). Driving forces emanate from aggregate effects of economy, technology, and demography. In this context, ideology adapts people to their surroundings and occurs largely in response to conditions created by material forces.

- In post-industrial society, post-modernization theory proposes a shift from materialist values (rooted in basic subsistence needs) to post-materialist needs (rooted in higher level needs and focused on quality of life issues). Forces of economic production change living conditions, while change in living conditions reprioritizes prevalent needs and values.

Management Implications

A cross-cultural perspective is critical for wildlife management professionals. Increasingly, conservation efforts are becoming multinational, professionals from developed countries are getting involved in developing country conservation issues, and government management agencies are considering multi-cultural perspectives. Analysis of the cultural context facilitates effectiveness in these situations.

Understanding and empowering people through new approaches to conservation. Cross-cultural conservation efforts increasingly apply concepts about culture. While introducing his volume on cross-cultural perspectives on wildlife in Asia, John Knight (2004) suggested that anthropology can describe local communities living at the wildlife interface; this provides assistance with *participatory conservation*, which is becoming conservation's new orthodoxy. The descriptive powers of anthropology also allow a critical examination of the wildlife profession, including the assumptions within management and conservation.

Anticipating trends. Understanding human–wildlife relationships in a broad cultural context helps management consider these issues as part of a long-term trend. For example, the recent trends against hierarchical decision making in natural resources is related to shifting needs and values and a changing lifestyle caused by changing modes of economy. This does not appear to be a short-term fad or a quickly reversible occurrence.

Attaining a deeper understanding of management issues. Analysis of management issues is enhanced when it integrates across scales. Individuals must be placed in their cultural context in order for situations to be thoroughly explained. For example, the current distribution of attitudes toward a management issue might be linked to more basic causes (power distribution, social structure, emerging economic forces, etc.). Solutions should target these basic causes (e.g., develop new modes of policy-making), instead of focusing on symptoms, i.e., specific issues.

Stimulating innovations to achieve sustainability. Solving tough human–wildlife sustainability issues (such as global warming) will depend on our ability to affect cross-cultural human thought and action. Strategies of change should consider ideology, which theoretically affects and is affected by culture shift.

References

Abel, T., & Stepp, J. R. (2003). A new ecosystems ecology for anthropology. *Conservation Ecology, 7*(3), 12.

Abramson, P. R. (1997). Postmaterialism and environmentalism: A comment on an analysis and a reappraisal. *Social Science Quarterly, 78*, 21–23.

Bell, D. (1973). *The coming of post-industrial society: A venture in social forecasting*. New York: Basic Books.

Bird-David, N. (1990). The giving environment: Another perspective on the system of gatherers-hunters. *Current Anthropology, 31*, 189–196.

Bird-David, N. (1992). Beyond the "original affluent society": A culturalist reformulation, *Current Anthropology, 33*, 25–47.

Boas, F. (1932). The aims of anthropological research. *Science, 76*(1983), 605–613.

Brechin, S. R., & Kempton, W. (1994). Global environmentalism: A challenge to the postmaterialism thesis. *Social Science Quarterly, 75*, 245–269.

Buckley, W. (1967). *Sociology and modern systems theory*. Englewood Cliffs, NJ: Prentice-Hall.

Bulmer, R. (1967). Why is the cassowary not a bird? A problem of zoological taxonomy among the Karam of the New Guinea Highlands. *Man, 2*, 5–25.

Burch, E. S. Jr., & Ellanna, L. J. (Eds.). (1994). *Key issues in hunter-gatherer research*. Oxford: Berg Publishers.

Butzer, K. (1989). Cultural ecology. In G. Gaile, & C. Willmott (Eds.), *Geography in America* (pp. 192–208). Merrill: Columbus.

Callicott, J. B., & mes, R. T. (Eds.). (1989). *Nature in Asian traditions of thought: Essays in environmental philosophy*. New York: Suny.

Carniero, R. L. (2003). *Volutionism in cultural anthropology: A critical history*. Boulder, CO: Westview Press.

Decker, D. J., Brown, T. L., & Siemer, W. F. (2001). *Human dimensions of wildlife management in North America*. Bethesda, MD: The Wildlife Society.

Deen, M. Y. I. (2001). Islamic environmental ethics, law, and society. In L. P. Pojman (Ed.), *Environmental ethics* (3rd ed., pp. 260–265). Belmont, CA: Wadsworth/Thomson Learning. (Reprinted from *Ethics of environment and development: Global challenges and international response*, pp. 190–196, by J. R. Engel, & J. G. Engel, (Eds.). (1990). London: Bellhaven Press.)

De Silva, L. (2001). The Buddhist attitude towards nature. In L. P. Pojman (Ed.), *Environmental ethics* (3rd ed., pp. 256–260). Belmont, CA: Wadsworth/Thomson Learning. (Reprinted from *Buddhist perspectives on the ecocrisis*, pp. 9–29, by K. Sandell, (Ed.). (1987). Kandy, Sri Lanka: Buddhist Publication Society.).

Douglas, M. (1957). Animals in Lele religious symbolism. *Africa, 27*(1), 46–58.

Douglas, M. (1990). The pangolin revisited: a new approach to animal symbolism. In R. Willis (Ed.), *Signifying animals: Human meaning in the natural world* (pp. 25–36). New York: Routledge.

Douglas, M., & Wildavsky, A. (1982). *Risk and culture: An essay on the selection of technical and environmental dangers.* Berkley: University of California Press.

Durkheim, E. (1964). *The elementary forms of religious life* (J. Swain, Trans.). New York: Free Press. (Original work published 1912).

Dwitvedi, O. P. (2001). Satyagraha for conservation: Awakening of the spirit of Hinduism. In L. P. Pojman (Ed.), *Environmental ethics* (3rd ed., pp. 260–265). Belmont, CA: Wadsworth/Thomson Learning. (Reprinted from *Ethics of environment and development: Global challenges and international response*, pp. 201–212, by J. R. Engel, & J. G. Engel, (Eds.). (1992). London: Bellhaven Press.)

Ferguson, R. B. (1995). Infrastructural determinism. In M. F. Murphy, & M, L, Margolis (Eds.), *Science, materialism, and the study of culture* (pp. 21–38). Gainsville: University Press of Florida.

Galvin, K. A., Thorton, P. K., Roque de Pinho, J., Sunderland, J., & Boone, R. B. (2006). Integrated modeling and its potential for resolving conflicts between conservation and people in the rangelands of East Africa. *Human Ecology, 34* (2), 155–183.

Gellner, E. (1988). *Plough, sword and book: The structure of human history.* Chicago: University of Chicago Press.

Harris, M. (1974). *Cows, pigs, wars and witches.* New York: Vintage.

Harris, M. (1979). *Cultural materialism: The struggle for a science of culture.* New York: Random House.

Harris, M. (1999). *Theories of culture in postmodern times.* Walnut Creek, CA: Altamira Press.

Herzfeld, M. (2001). *Anthropology: Theoretical practice in culture and society.* Malden, MA: Blackwell Publishers.

Inglehart, R. (1990). *Culture shift in advanced industrial societies.* New Jersey: Princeton University Press.

Inglehart, R. (1997). *Modernization and postmodernization.* New Jersey: Princeton University Press.

Inglehart, R., & Baker, W. E. (2000). Modernization, cultural change, and the persistence of traditional values. *American Sociological Review, 65,* 19–51.

Inglehart, R., & Welzel, C. (2005). *Modernization, cultural change and democracy: The human development sequence.* New York: Cambridge University Press.

Ingold, T. (1994). From trust to domination: An alternative history of human-animal relations. In A. Manning, & J. Serpell (Eds.), *Animals and human society* (pp. 1–22). New York: Routledge.

Kellert, S. R. (1993). Attitudes, knowledge, and behavior toward wildlife among the industrial superpowers: United States, Japan, and Germany. *Journal of Social Issues, 49,* 53–69.

Knight, J. (Ed.). (2004). *Wildlife in Asia.* Curzon, London: Rouledge.

Leach, E. R. (1964). Anthropological aspects of language: Animal categories and verbal abuse. In E. H. Lenneberg (Ed.), *New directions in the study of language* (pp. 23–63). Cambridge, MA: MIT Press.

Leach, E. (1970). *Claude Lévi-Strauss.* Chicago: University of Chicago Press.

Lévi-Strauss, C. (1963). *Totemism* (R. Needham, Trans.). Boston: Beacon.

Malinowski, B. (1939, May). The group and the individual in functional analysis. *The American Journal of Sociology, 44*(6), 938–964.

Mendelsohn, R., & Sachs, J. (2006, September 9). Dismal calculations. *The Economist*, 380 (8494). Retrieved from http://www.economist.com/specialreports/displaystory.cfm?story_ id = 7853042.

Millennium Ecosystem Assessments. (2005). *Ecosystems & human well-being: synthesis*. Washington, DC: Island Press.

Milton, K. (1996). *Environmentalism and cultural theory*. London: Routledge.

Mithen, S. (1996). *The prehistory of the mind: The cognitive origins of art, religion and science*. New York: Thames and Hudson.

Mullin, M. H. (1999). Mirrors and windows: sociocultural studies of human-animal relationships. *Annual Review of Anthropology, 28*, 201–204.

Murphy, M. F., & Margolis, M. L. (1995). An introduction to cultural materialism. In M. F. Murphy, & M. L. Margolis (Eds.), *Science, materialism, and the study of culture* (pp. 1–4). Gainsville: University Press of Florida.

Osborne, A. (1990). Eat and be eaten: animals in U'wa (Tunebo) oral tradition. In R. Willis (Ed.), *Signifying animals: Human meaning in the natural world* (pp. 140–158). New York: Routledge.

Pálsson, G. (1990). The idea of fish: land and sea in the Icelandic world-view. In R. Willis (Ed.), *Signifying animals: human meaning in the natural world* (pp. 119–133). London: Routledge.

Radcliffe-Brown, A. R. (1952). *Structure and function in primitive society*. London: Cohen and West.

Richerson, P. J., & Boyd, R. (2006). *Not by genes alone: how culture transformed human evolution*. University Of Chicago Press.

Robbins, P. (2004). *Political ecology: critical introductions to geography*. Malden, Mass: Blackwell.

Salzman, P. C. (2001). *Understanding culture: An introduction to anthropological theory*. Prospect Heights, IL: Waveland Press.

Schwabe, C. (1994). Animals in the ancient world. In A. Manning, & J. Serpell (Eds.), *Animals and human society: Changing perspectives* (pp. 36–58). London: Routledge.

Schwartz, S. H. (2006). A theory of cultural value orientations: Explication and applications. *Comparative Sociology, 5*, 136–182.

Schwartz, S. H., & Sagie, G. (2000). Value consensus and importance: A cross-national study. *Journal of Cross-Cultural Psychology, 31*(4), 465–497.

Scoones, I. (1999). New ecology and the social sciences: what prospects for a fruitful engagement? *Annual Review of Anthropology, 28*, 479–507.

Shanklin, E. (1985). Sustenance and symbol: Anthropological studies of domesticated animals. *Annual Review of Anthropology, 14*, 375–403.

Simoons, F. J. (1979). Questions in the sacred cow controversy. *Current Anthropology, 20* (3), 476–476.

Steward, J. (1955). *Theory of cultural change: The methodology of multilinear evolution*. Urbana: University of Illinois Press.

Sukumar, R. (1989). *The Asian elephant: ecology and management*. Cambridge: Cambridge University Press.

Swingewood, A. (1998). *Cultural theory and the problem of modernity*. Hamburg, Hong Kong: Macmillan Press.

Theodossopoulos, D. (2005). Care, order and usefulness: The context of the human-wildlife relationship in a Greek island community. In J. Knight (Ed.), *Animals in person: Cultural perspectives on human-animal intimacy* (pp. 15–36). New York: Berg.

Trigger, B. G. (1998). *Sociocultural evolution: Calculation and contingency*. Blackwell, Malden Mass.

Weber, M. (1948). *The Protestant ethic and the spirit of capitalism*. (T. Parsons, Trans.). New York: Scribner. (Original work published 1930).

White, L. (1953). *The evolution of culture*. New York: McGraw-Hill.

White, L. (1967). The historical roots of our ecologic crisis. *Science, 155*, 1203–1207.

Wildavsky, A. B. (1991). *The rise of radical Egalitarianism*. Washington, DC: The American University Press.

Willis, R. (1990). Introduction. In R. Willis (Ed.), *Signifying animals: human meaning in the natural world* (pp. 1–24). London: Routledge.

Wolf, E. (1972). Ownership and political ecology. *Anthropological Quarterly, 45*, 201–205.

Chapter 8
Integrating Concepts: Demonstration of a Multilevel Model for Exploring the Rise of Mutualism Value Orientations in Post-industrial Society[1]

Contents

Introduction

In the early 1990s, I conducted a survey of public values toward wildlife for the Colorado Division of Wildlife. The purpose of the study was to inform statewide wildlife planning. The planning was being conducted, in part, to identify new agency directions and embrace new stakeholders for wildlife in Colorado. The survey we conducted was one of our first uses of the concept of wildlife value orientations (see Bright, Manfredo, & Fulton, 2000; Fulton, Manfredo, &

---

[1] This chapter was co-authored with Tara L. Teel.

M.J. Manfredo, *Who Cares About Wildlife?*,
DOI: 10.1007/978-0-387-77040-6_8, © Springer Science+Business Media, LLC 2008

Lipscomb, 1996). After completion of the study, I was invited to report the findings at a formal meeting of the Colorado Wildlife Commission, the group with regulatory authority over wildlife in the state. That small group of people, appointed by the governor, is charged with representing the interests of the public in making decisions about wildlife in Colorado. Our findings showed that about three in ten Coloradoans held pro-animal-rights and antihunting beliefs. These results surprised members of the commission. After listening to my presentation, the chair of the Commission proceeded to deride the study findings. "I know the Colorado public," he exhorted in a loud, angry voice, "and this is not what they are like. People in Colorado hunt and love hunting." There were others who, in a more subdued fashion, acknowledged the significance of the findings. The display at that meeting, for me, is representative of a pervasive tension in North American wildlife management today. Professionals are uncertain about how to deal with growing societal opposition to the techniques and traditions of wildlife management. Controversy is most obvious on issues of hunting. For some professionals, the decline of hunting participation and the perceived rise of antihunting sentiment prompt responses of denial (such as I received), the urge to "fight back" against those who oppose hunting (e.g., passage of hunter rights bills – see Jacobson, 2006), and efforts directed at turning the trend of declining numbers of hunters (e.g., there is a proliferation of state agency hunter recruitment and retention efforts). At the same time, there are those within the wildlife profession who believe we must better understand and adapt to changing societal expectations about wildlife management. The case study reported here was commissioned by the Western Association of Fish in Wildlife Agencies in the spirit of learning and understanding. The purpose of the study was to examine the possibility of wildlife value shift among publics in the western United States, its causes, and its effects on the future of wildlife management.

Strengthening the Conceptual Foundation of HWD Research

As global economies and human populations grow, as development expands, and as global warming takes its toll, social conflict over a limited natural resource base will inevitably increase in the twenty-first century. The need to understand the human phenomena surrounding this conflict will also increase, as will the need for social science information that can guide conservation action. As the ramifications of conflict intensify, human dimensions of wildlife (HDW) researchers will be faced with the challenge of improving the utility of their work. The premise of this book is that the utility of research can be improved by strengthening the conceptual foundation of investigations. This will enhance our ability to understand human behavior, predict its occurrence, and affect it in achieving conservation goals.

To contribute to improved applications in HDW research, this book centers around two main goals. One is to provide an update and overview of theoretical concepts commonly used in the study of human–wildlife relationships. Currently, that would include attitudes, norms, and values (Manfredo, Teel, & Bright, 2004; Vaske, Shelby, & Manfredo, 2006). Keeping abreast of, and participating in the creation of theoretical advancements in, the attitudes, norms, and values literature will ultimately improve application of these concepts. Another goal of this book is to offer insight into new conceptual directions for HDW research. Examples of potentially promising topics (but certainly not an exhaustive list) include dual-processing approaches to understanding attitudes and attitude change, cross-cultural values and ideology, moral norms and emotions, the role of sanctions in directing normative behavior, and the role of social identity in forming attitudes toward wildlife issues. In particular, topics that have been generally neglected in HDW research include emotions and the heritability of responses to wildlife. Of these two, work in the area of emotions and mood would have a high degree of applied relevance. The role of emotions in persuasion, stakeholder processes, and employee satisfaction and effectiveness seems particularly relevant. Research might also, for example, integrate the concepts of values, norms, and emotions in understanding social roles and social identities related to conservation.

Another very important consideration for future research is the need to integrate across the social science disciplines. The importance of integrating the social and ecological sciences is widely recognized; however, bridging the gaps among the various social science disciplines is equally important. Social phenomena are the result of complex interactions of influences at multiple scales, and different disciplines contribute uniquely in this context. It will therefore be critical to consider multiple social science perspectives in order to account for interactions among such diverse influences as material forces (economic, technological, and demographic factors), social-structural variables, institutions, social organization, social groups, and individual-level cognitions and emotions.

Example Case Study

For the purpose of illustrating new research directions, the remainder of this chapter provides a case study demonstration. The case study stems from cooperative research among the state wildlife agencies of the Western Association of Fish and Wildlife Agencies (WAFWA). Cooperation allowed for comparisons across states (macro level) and led to conclusions about the effects of a modernizing society on wildlife value orientations. Specifically, our thesis proposes the re-emergence of a mutualism orientation in post-modern societies (Manfredo, Teel, & Henry, 2007). In addition to its conceptual contribution, this case study illustrates an emerging type of statistical technique that will be useful for examining multilevel social phenomena.

Study Background: The Re-emergence of a Mutualism Wildlife Value Orientation Study Question

Research shows a worldwide twentieth-century trend in which societal values have become more self-expressive (Inglehart, 1997; Schwartz, 2006) and attitudes more pro-environmental (Dunlap, 2002). Have wildlife value orientations also changed?

Evidence from North America suggests that such change is possible. Trends such as the decline of recreational hunting in North America and the mobilization of non-governmental agencies advocating animal rights suggest that societal views toward wildlife are changing (Heberlein, 1991; Muth & Jamison, 2000; Peyton, 2000; U.S. Fish and Wildlife Service, 2007). A trend toward increased social conflict over wildlife management issues seems to support the emergence of new interests. Minnis (1998), for example, reported that, prior to 1972, there was just one antihunting/anti-trapping ballot initiative (banning trapping in Massachusetts). In the 1970s and 1980s, five initiatives were brought to the ballot, and only one passed. In the 1990s, however, 14 initiatives were brought forward, of which nine passed.

Other evidence comes from changes within the wildlife profession. University programs that offer wildlife management degrees, for example, have witnessed a change in the characteristics of students attracted to their programs. Increasingly, these students do not hunt and, in fact, to a greater extent, now have antihunting attitudes (Organ & Fritzell, 2000). This is consistent with research that notes declines in hunting participation among wildlife professionals, along with a slight shift away from utilitarian wildlife values (Brown, Connelly, & Decker, 2006).

Given the lack of longitudinal data, it is not surprising that there are few studies that directly examine wildlife value shift. Those available, however, suggest a change has occurred. Kellert (1976), through an analysis of American newspaper accounts between 1900 and 1976, found a decrease in utilitarian attitudes toward wildlife. Similarly, Manfredo and Zinn (1996) speculated about a trend away from utilitarian value orientations based on an analysis of intergenerational differences in a sample of Colorado residents.

The purpose of the case study reported here was to examine the possibility of shifting wildlife value orientations. The study asked whether wildlife value orientations are changing in North America and, if so, what societal-level factors might be contributing to this occurrence.

Central Thesis: The Re-emergence of a Mutualism Orientation

Anthropologists have long observed a linkage between human–wildlife relationships and human–human relationships within societies (see Chapter 7). Ingold (1994), for example, proposed that with the transition from hunter-gatherer to agriculturally based societies, there was a shift from trust and mutualism to domination in human–human relationships *and* in human–wildlife relationships. The case study reported here proposed the return to trust and mutualism views toward wildlife in post-industrial societies. Following Ingold's logic, such a shift is likely, given the changing nature of interpersonal human relationships in these societies. There has been a decline in concern for material well-being (Inglehart, 1997), a rise in egalitarian relations (Wildavsky, 1991), and a growing concern for belongingness and self-esteem (Inglehart and Welzel, 2005). These trends in human–human relationships set the stage for a re-emergence of trust views toward wildlife and a decline in views of dominance and mastery.

To test that proposal, a micro–macro model was developed (Manfredo et al., 2007) that explored several key questions. First, what forces of modernization have affected change in societal values and wildlife value orientations? Second, how do these forces affect the circumstances of daily life in a way that would stimulate value orientation shift? Third, how do value orientations and such a shift affect the observable behavior of individuals? The micro and macro components of the model are discussed separately below.

Micro Model: Wildlife Value Orientations and Individual Action

The general structure of the micro model was the well-established cognitive hierarchy that has been used in several past studies in HDW (see Chapter 6). What is unique about the application described here is the explicit integration of ideology into this framework using the wildlife value orientation concept. After a brief overview of the cognitive hierarchy, a lengthier discussion is provided below on wildlife value orientations.

Cognitive hierarchy. The cognitive hierarchy, or value–attitude–behavior framework (VAB), integrates topics reviewed in previous chapters including values, attitudes, norms, and behaviors (Homer & Kahle, 1988; Maio Olson, Bernard, & Luke, 2003). According to this framework, attitudes and norms are the proximate causes of wildlife-related behaviors that include, for example, recreational hunting, feeding wildlife in one's yard, caring for abandoned wildlife, voting on wildlife-related ballot initiatives, and watching wildlife-related TV programs. While attitudes and norms influence behavior, they are ultimately linked to more basic and overarching cognitions referred to as values. We adopted Inglehart's (1997) approach of classifying values in our empirical tests primarily because of Inglehart's explicit theoretical link to macro explanations of change. He uses a classification scheme consisting of materialist and post-materialist values for describing global value shift in post-modern societies. Materialist values prioritize goals (e.g., fighting crime, maintaining a stable economy) that originate in basic needs, such as safety and survival, whereas post-materialist values prioritize goals (e.g., giving people more say in government, protecting the environment, ensuring freedom of speech) that originate in needs related to belongingness, self-esteem, and self-fulfillment.

Wildlife value orientations: ideology in the context of VAB. The notion of value orientations explicitly accounts for the influence of ideology on the structure of cognitions. These orientations are reflected in dimensions such as mastery versus harmony (Schwartz, 2006), communal sharing versus authority ranking (Fiske, 1992), and individualism versus collectivism (Triandis, 1995). The integration of ideology into the VAB framework is critical in understanding the meaning assigned to values. The ideological orientation of values, for example, will help us distinguish between the different attitudes and behaviors of two individuals who assign high priority to a value such as *protecting the welfare of nature* (Schwartz, 2004). One person may believe that this means people should not harm wildlife for any reason, while the other may feel it is acceptable to kill wildlife for food if one ensures the animal does not experience unusual pain and suffering. These two individuals would act differently toward wildlife, yet purport to hold the same value.

Recent usages of the term *value orientations* focus on value groupings and value priorities (e.g., Rokeach, 1973); however, these applications vary somewhat from the original concept of value orientations introduced by Kluckholn (1951). He proposed that value orientations represent, at both the individual and group levels, *unity thema* or ethos that capture the personality of a cultural group. A value orientation, according to Kluckholn (1951, p. 411), is "...a generalized and organized conception, influencing behavior, of nature, of man's place in it, of man's relation to man, and of the desirable and non-desirable as they may relate to man–environment and inter-human relations." His concept of value orientations was illustrated in a study of groups in the American Southwest where he determined that Mormons had orientations described as *mastery over nature*, Spanish-Americans had *subjugation* orientations, and Navaho had *harmony with nature* orientations. These orientations were proposed to have a profound effect on how one sees the world and interacts with it.

Our notion of wildlife value orientations is similar to Kluckholn's conceptualization.

Wildlife Value Orientations Reflect the Infusion of Broad-Based Ideology into the VAB Hierarchy

Ideology orients the pattern, strength, and direction of beliefs that form around a value and give the value-specific contextual meaning. For example, a value such as *being humane* may be expressed in several schematic domains including *dealing with other people, helping those in need*, and *treatment of other living things*. These separate, domain-specific schema, however, form a thread of consistency through the influence of ideology. To illustrate, an *egalitarian* ideology would be revealed in beliefs that suggest the importance of equality in treatment of others, being charitable toward those in need, and caring and extending trust toward other forms of life.

The strength of a given ideology, and hence value orientations, varies among individuals (Pratto, 1999; Sidanius, 1993), and differences in attitudes, norms, and behaviors are created from this variation. The nature of this influence, as it relates to wildlife, is described through two primary value orientations: *domination* and *mutualism*.

Domination wildlife value orientation. Kluckholn and Strodbeck (1961) proclaimed that domination is the primary orientation of most Americans, and contemporary empirical findings reinforce that conclusion (Schwartz, 2006). The domination ideological orientation has been tied to the rise of Judeo-Christian religion, the extensive worldwide colonization that emanated from European countries during the last millennia, the emergence of science and technology, the global expansion of capitalism, and the rapid growth of environmental degradation that has become salient in recent history (Buttel & Humphrey, 2002; Catton & Dunlap, 1980; Hand & Van Liere, 1984; Werner, Brown, & Altman, 1997; White, 1967). Schwartz's (1994, 2006) cross-cultural studies of values support the notion that domination (labeled *mastery*) is a widespread cultural dimension that lies in opposition to a harmony with nature orientation.

Social domination theory (SDT), developed to explain hierarchies in human society, applies readily to explanations of human–wildlife relationships. As introduced, SDT focuses on the resultant prejudice, discrimination, and stereotyping that is associated with the human tendency to form group-based hierarchies (Pratto, 1999; Sidanius, 1993; Sidanius, Pratto, van Laar, & Levin, 2004). For societies with economic surplus, group-based hierarchies form in social institutions and powerful individuals direct desirable items to members of dominant and privileged groups while directing undesirable things to less powerful groups. Ideology justifies and is reinforced by power differentials and the social, legal, and economic practices that create them. That is, the order of dominance is sustained through the way social practices, belief systems, and psychological processes are linked in shared ideologies (Pratto, 1999).

Social domination theory implies that the transition from trust to domination in human–animal relationships facilitated humans assuming power over animals. This transition increased benefits to humans and, in the process, relegated animals to a group receiving undesirable roles and conditions (e.g., heavy work loads, poor living conditions, expendable lives for human purposes). As the theory predicts, the domination ideology results in a clear separation of groups (i.e., animals from humans), stereotypes that reinforce roles and maintain purposes of human advantage (e.g., good animals and bad animals), beliefs that provide justification for cultural practices (e.g., beliefs that animals exist to advance the needs of humans or that animals have no feelings,

emotions, or capacity for culture), and mythology that emphasize the subordi-
nate role of animals to humans (e.g., *Genesis* 1:28).

To summarize, a *domination wildlife value orientation* reflects the extent to
which an individual (or group) holds an ideological view of human mastery of
wildlife. It is expected that the stronger a person's domination orientation, the
more likely her attitudes and actions will prioritize human well-being over
wildlife. She will find actions that result in death or other intrusive control of
wildlife to be acceptable, and she will find justification for treatment of wildlife
in utilitarian terms.

Mutualism wildlife value orientation. A mutualism wildlife value orientation
reflects the influence of egalitarian ideology in the VAB hierarchy. Schwartz
(2006) finds support for the presence of an egalitarian ideology that varies
across countries. Wildavsky (1991) argued that the growth of animal rights in
America originated in the rise of *egalitarian culture.* An egalitarian ideology
places emphasis on equality and on individuals acting for the welfare of all. *All,*
according to the animal rights movement, includes both humans and animals
because the two share equal moral status.

The egalitarian ideology leads to the social inclusion of animals, and the
human–animal relationships that arise from this view are shaped by (a) the
motivational forces borne from a need for belongingness and (b) the human
tendency to anthropomorphize. The need for belongingness and affiliation is
recognized as a human universal. Maslow (1954), for example, proposed that
this need was prominent after basic existence needs were met. In a review of
literature on this topic, Baumeister and Leary (1995) concluded belongingness
is a biologically prepared human trait that directs people to seek contacts and
interactions with other persons and to establish bonds that are perceived to be
stable and affectively rewarding. Moreover, the failure to meet belongingness
needs is linked to pathological conditions such as increased stress or negative
psychological and physical health conditions.

As noted by Baumeister and Leary (1995), satiation of belongingness needs
generally requires perceived mutuality. That is, for those seeking mutualism
bonds with wildlife, it should be perceived that the animal returns caring,
gratitude, or a similar response. This is facilitated by the strong tendency of
humans to anthropomorphize. The tendency to project human characteristics
on wildlife and domesticated animals is a universal human trait that emerged
due to the evolutionary advantages it posed to hunters and gatherers (Mithen,
1996). In post-industrialized society, however, the anthropomorphic tendency
is shaped in a way that promotes perceptions of a sense of social connectedness
with, and caring for, animals (Katcher & Wilkins, 1993; Serpell, 2003; Vining,
2003). As is discussed in the macro section of this chapter, the post-industrial
tendency for humans to project human characteristics that invoke affiliation
responses has been enhanced by (a) lifestyles in which there is a physical
removal of people from wildlife, (b) the increasing decline of an association
between wildlife and material needs (i.e., wildlife as a food source), and (c) the
pervasive media portrayal of wildlife as human personalities.

In summary, the *mutualism wildlife value orientation* shaped from these influences views wildlife as capable of living in relationships of trust with humans, as life forms having rights like humans, as part of an extended family, and as deserving care and compassion. Those with a strong mutualism orientation would be more likely to engage in welfare-enhancing behaviors for individual wildlife (e.g., feeding, nurturing abandoned or hurt animals), and less likely to support actions resulting in death or harm to wildlife. Mutualists are more likely to view wildlife in human terms, with personalities and characteristics like humans.

The nature of the mutualism orientation in post-modern society only slightly resembles the mutualism view prevalent among hunter-gatherer societies. Milton (1996) indicated that modern-day environmentalists project a romantic view of hunters and gatherers as a people who lived in harmony with the environment and in trust with other animals. This view, she suggested, may serve as a powerful myth in galvanizing environmentalist groups; however, the notion of trust in hunter-gatherer societies was created from a different cultural context. Beliefs were formed within an ideology of subjugation, where people believed they were at the mercy of natural and supernatural forces to which they paid homage. However, the mutualism orientation of post-modern society is linked to an egalitarian ideology that exists in a broader context of domination. For example, caring for animals and the environment in post-modern society is not based on fear of reprisal from supernatural forces. Instead, it is based on beliefs that humans have the power and responsibility to act. In this view, humans believe they have the ability to protect animals, and can decide either to protect animals or not to protect them, as opposed to protecting animals for fear of punishment from supernatural forces.

In summary of the micro model, individual behavior toward wildlife is motivated by specific attitudes, and these attitudes are directed by wildlife value orientations. Wildlife value orientations are reflected in ideologically shaped beliefs that give personal meaning of right and wrong in human–wildlife relationships. Wildlife value orientations play an important role in explaining variation in people's wildlife-related behaviors and their attitudes toward topics related to wildlife treatment.

Measurement of Wildlife Value Orientations

In prior research, two approaches have been used to measure wildlife value orientations. A qualitative approach, used in cross-cultural studies (see *Human Dimensions of Wildlife*, Volume 12, Issue 1), asks participants to recount stories about wildlife that evoke certain emotional responses (e.g., a story about wildlife that made them sad). For a more detailed description of this methodology, the reader is directed to Dayer, Stinchfield, and Manfredo (2007). The second approach, applied in the case study reported here, is quantitative and was developed for use specifically in North America. It uses item scales whose development was guided by psychometric considerations of validity and reliability. Readers interested in the details of the quantitative assessment procedure are directed to Appendix.

Source of Data

For purposes of this chapter, a brief summary of study methods is provided here. See Teel et al. (2005) for a more detailed description. The study was conducted in 19 states in the western United States in cooperation with WAFWA. Data were collected via a mail survey administered to a sample of residents in each state in October and November 2004. There were 12,673 completed surveys – over 400 for each state. A total of 69,031 surveys were mailed out, and 8,063 surveys were returned by the U.S. Postal Service as nondeliverable, resulting in an overall response rate of 21%. A telephone nonresponse check ($n = 7,388$) revealed some differences, and data were weighted as a result by age and wildlife-associated recreation participation for reporting of population estimates.

In addition to measuring wildlife value orientations, the survey assessed (a) values, using Inglehart's items, (b) attitudes toward wildlife management issues (one set of items was identical across all states, and one set was unique to each state and contained in a separate state-specific section of the survey), (c) participation in wildlife-associated recreation, and (d) sociodemographic characteristics.

Tests of the Micro Model

The traditional VAB framework proposes that values affect attitudes, and attitudes, in turn, affect behavior. Our theory proposes that at the micro, or individual, level, the influence of values on attitudes is directed by the influence of value orientations. Two tests of this model are summarized here. The first computed correlations of wildlife value orientations with the many attitudinal variables assessed across state-specific versions of the survey (see Teel & Manfredo, 2007a). Overall, this analysis showed statistical significance for 71% of all correlations with the domination orientation (337 out of 473 attitude items; 107 with correlations of 0.30 or greater). For mutualism, correlations were statistically significant for 59% of all attitudinal measures (279 items, 87 with correlations of 0.30 or greater). Correlations were stronger for items involving management issues with implications of direct or indirect harm to wildlife. It is reasonable to deduce from this analysis that the mutualism and domination value orientations had strong predictive validity in this study.

The second test examined the entire model, i.e., values→value orientations→attitudes/behaviors. This required a more complicated analysis procedure to test whether wildlife value orientations statistically mediated the value→attitude/behavior relationship. Two separate dependent measures were included in this analysis: (a) attitudes toward the management action, *providing more recreational opportunities to hunt* black bears when the bears enter residential areas and create a nuisance (i.e., bears getting into trash and pet food containers; 0 = *unacceptable*, 1 = *acceptable*); and (b) hunting participation (behavior) within the past 12 months (*yes/no*). According to criteria outlined by Baron and Kenny (1986), mediation would be established for this model if (a) values affect the dependent measures (attitudes/behavior); (b) values affect value orientations; and (c) value orientations affect the dependent measures when controlling for the effect of values. Partial mediation is demonstrated if all criteria are met. Full mediation is established if all criteria are met *and* values have no direct effect on the dependent measures once value orientations are included in the model.

Results, using structural equation modeling with probit regression, showed that the domination wildlife value orientation fully mediated the relationship between values and attitudes toward lethal control of bears (Table 8.1). All other relationships displayed partial mediation. As indicated by the direction of coefficients, results indicated that those with a stronger domination orientation expressed greater support for lethal control of bears. Domination was also positively associated with participation in hunting and with Inglehart's materialist values set. In contrast, as expected, mutualism was negatively associated with support for lethal bear management and hunting and positively associated with post-materialist values.

These findings provide strong empirical support for the micro model. That is, they provide support for the idea that values affect wildlife-related attitudes and behaviors indirectly through wildlife value orientations. The latter are key to explaining why people with similar societal values can end up on opposite sides of wildlife management issues and display very different behaviors toward wildlife. Given this micro-level phenomenon of wildlife value orientations forming the basis for differences in societal thought and action toward wildlife, an important question becomes, what factors affect the presence of a domination versus a mutualism orientation in society? This brings us to the macro component of our model.

Macro Model: Exploring the Impact of Modernization on Wildlife Value Orientations

As reviewed in Chapter 7, the forces of modernization have been central to theories of societal and cultural value shift. Common to these theories is a view that cultural thought is formed either directly or indirectly in response to

Table 8.1 Results of mediation analysis to assess the role of wildlife value orientations in the VAB framework

	Estimate[a]	SE[b]	t[c]
Values → Domination → Attitudes toward lethal bear control	–	–	–
Values → Value orientations	−0.06	0.00	−17.23
Value orientations → Attitudes/behaviors	0.56	0.01	64.37
Values → Attitudes/behaviors (direct effect)[d]	0.00	0.00	−0.40 (NS)
Values → Attitudes/behaviors (indirect effect)	−0.03	0.00	−18.02
Values → Domination → Hunting	–	–	–
Values → Value orientations	−0.06	0.00	−17.23
Value Orientations → Attitudes/behaviors	0.47	0.01	42.87
Values → Attitudes/behaviors (direct effect)	0.01	0.00	3.36
Values → Attitudes/behaviors (indirect effect)	−0.03	0.00	−14.78
Values → Mutualism → Attitudes toward lethal bear control	–	–	–
Values → Value orientations	0.05	0.00	14.88
Value orientations → Attitudes/behaviors	−0.31	0.01	−28.26
Values → Attitudes/behaviors (direct effect)	−0.02	0.00	−7.72
Values → Attitudes/behaviors (indirect effect)	−0.02	0.00	−12.36
Values → Mutualism → Hunting	–	–	–
Values → Value orientations	0.05	0.00	14.88
Value orientations → Attitudes/behaviors	−0.13	0.02	−8.53
Values → Attitudes/behaviors (direct effect)	−0.01	0.00	−3.50
Values → Attitudes/behaviors (indirect effect)	−0.01	0.00	−7.55

[a] Estimates represent probit regression coefficients computed in Mplus 4.2 (Muthén & Muthén 1998–2006).
[b] SE = Standard error associated with estimates (regression coefficients).
[c] Unless otherwise noted, t value is significant at $p < .05$. NS = non-significant at this alpha level.
[d] Direct effect of values after adjusting for the effect of the mediator (value orientations) on attitudes/behaviors.

material conditions (e.g., mode of economy, demography, technology, environment). Inglehart (1990, 1997; Inglehart & Welzel, 2005), for example, proposed a shift from traditional/religious values during pre-industrial periods to secular/rational values during industrial periods to emphasize on self-expression values in post-industrial periods. Change, Inglehart proposed, is driven by technological innovation that leads to increased economic productivity. With that comes occupational specialization and rising education and income levels, which in turn change the nature of daily life. Value shift occurs as economic productivity lessens the importance of subsistence needs and elevates the importance of self-expression needs.

Changing lifestyles and wildlife value orientations. The lifestyle arising from modernization in North America has altered the nature of human interactions with wildlife. As described earlier in the chapter, a domination wildlife value orientation, with emphasis on basic material needs and utilitarian concerns, is linked to the colonization and westward expansion of North America. A strong tradition of hunting emerged during this time period (Cochrane, 1993; Muth & Jamison, 2000), and achievement in hunting became highly symbolic of manhood, social status, and rural life (Dizard, 2003; Herman, 2003).

The processes of modernization, including economic growth, urbanization, and rising education level transformed human–wildlife relationships by changing the need structure of society and the day-to-day environments that people experienced. The transition from industrialization to post-industrialization brought changes in technology and economic production that created new life conditions for most Americans. They had higher income and education, were more likely to reside in urban areas, and were increasingly employed in the service industry. As highly efficient food-producing agricultural systems emerged and economic well-being extended to all social classes, the need for wildlife as a food source and the impetus for domination in wildlife relationships were greatly diminished. Theorists have proposed that with growing wealth and economic productivity, the prevalence of existence needs was removed making salient needs for belongingness, esteem, and quality of life (Dillman & Tremblay, 1977; Inglehart, 1997). This changed the perceptual basis for interaction with wildlife. The motivation to see wildlife as a contribution or as a threat to economic well-being was eliminated for most people. However, with the increased prominence of belongingness needs and the tendency to anthropomorphize, wildlife would be seen as potential companions and part of one's social group.

As the motivational basis of human–wildlife relationships changed, so did the context of people's experiences. Bell (1973) reflected these changes in his proposal that life was a "game against nature" in agrarian times, "a game against fabricated nature" during industrialization, and "a game against people" in post-industrial times. In this transition, interactions with wildlife changed from *competitive* to *benign, incidental, and recreational*, particularly for those of the modernized social classes (i.e., those with higher levels of education and income, and residing in urban areas). Daily encounters with wildlife in rapidly expanding urban and suburban areas became infrequent, and those that occurred were with species readily adaptable to human settlement, including some small mammals and birds. Increasingly, the context of interactions with wildlife involved minor nuisance situations or outdoor recreation where wildlife were perceived as objects of curiosity and learning.

As the frequency of direct contact with wildlife diminished, experiences were replaced with characterizations, formed through media and modern mythology, in which wildlife were blatantly portrayed as having human characteristics. By making human-like relationships with wildlife seem possible, this anthropomorphic tendency provided further impetus for a rise in mutualism to meet the emerging need for belongingness in post-industrial society. Furthermore, it contributed to the creation of a social environment much less tolerant of a domination view. Change, which was particularly obvious in relation to recreational hunting activities, was evident in the nature of behaviors directed toward wildlife (e.g., hunting participation diminished), in social interactions (e.g., the declining acceptability of hunting-related topics in social discourse), and in social traditions (e.g., diminished importance of a young man's first hunt, the prestige assigned to being a good hunter, and state-recognized holidays for hunting season).

Finally, it should be noted that this change has been slow and intergenerational. As indicated by Bell (1973), in the process of value orientation shift, one orientation is not replaced by another; rather, the multiple perspectives are overlaid upon one another, enriching the diversity and complexity of value orientations present in society.

Tests of the macro model. Our macro theory proposes that increasing modernization has led to the rise of a mutualism value orientation toward wildlife. To provide a specific test of that proposal, data from the 19 western United States were analyzed to determine whether there was a relationship between variables that would indicate modernization (income, urbanization, education) and the prevalence of a mutualism orientation. There are four important clarifications about this test.

First, the variables used in this analysis *implicate* modernization, but by themselves are not the proximate cause of value shift. Rather, they are believed to indicate a particular style of life that, as a whole, gives rise to a mutualism view within a society.

Second, the change we propose occurs at a group or societal level, not at the individual level. The nature of effects in group-level change is quite different from the forces of individual change. Values and value orientations are formed in one's youth and typically change little over the course of one's lifetime. A more modernized lifestyle in later life, therefore, is unlikely to change a person's value orientations. As an illustration, a person does not move to a city and suddenly adopt a mutualism orientation. Instead, urbanization is indicative of a more modern lifestyle that would affect the person's offspring, producing, in essence, a delayed and indirect effect through time. Hence, the change we are proposing would be apparent over multiple generations.

Third, the theory proposed here suggests a longitudinal process of change, but the actual test provided below is a cross-sectional analysis. The test presumes that if change has been occurring as proposed, we would see a particular pattern of findings. Specifically, we hypothesized that individuals residing in more modernized states (i.e., with higher levels of urbanization, education, and income) would be more likely to have a mutualism value orientation.

Fourth, the type of analysis necessary for this research problem must account for the unique problems of multilevel analysis. Prior approaches to examining relationships among societal-level and individual-level phenomena are often subject to a criticism referred to as the *ecological fallacy* (or *reverse ecological fallacy*). This occurs when relationships revealed at one level of analysis are generalized to another level (Hofstede, 2002). For example, analyses conducted at the individual level rarely show a strong correlation between urbanization, education, or income and value orientation measures. Those individual-level analyses should not be considered generalizable to relationships that might exist at the aggregate, or societal, level.

Analysis reported here used hierarchical linear modeling (HLM), which is recognized as the appropriate form of analysis for multilevel data (Raudenbush & Bryk, 2002). States were used as the cases at the macro level, and individuals within states were the unit of observation at the micro level. HLM allowed for simultaneous testing of individual and societal-level effects on an individual-level outcome, thereby providing a more thorough test of relationships between macro-level forces and individual thought or behavior. Results of this analysis, taken from Manfredo et al. (2007), are shown in Table 8.2 and support the proposal that modernization is related to the prevalence of a mutualism wildlife value orientation. In particular, results showed that an individual's score on both mutualism and domination is affected by the level of income, urbanization, and education in that person's state of residence. This is revealed by looking at coefficients in the column labeled *contextual effects* that should be interpreted as unstandardized regression coefficients. The coefficients indicate the change in an individual's score on the value orientation scale produced by a 1-unit increase in the independent variable (income, urbanization, education). So, for example, a 1-unit increase in a state's mean level of education produces a 0.92 increase in a person's score on the mutualism scale, while controlling for the effect of that person's own level of education. This might also be interpreted as follows: the coefficient (0.92) is the expected difference in mutualism scoring between two individuals who have the same education level but who reside in states differing by 1 unit in mean level of education. It can be concluded from this analysis that there is something about the state in which an individual resides (defined by the modernization variables) that has a significant impact on that individual's value orientations above and beyond any effect due to that individual's own level of wealth, education, or size of community. The results show that this is true even after controlling for variables such as age and gender.

Overall, how substantial is the effect? Results indicated that between-group means for the value orientations vary across states more than would be expected by chance alone, given levels of variation within states. Calculation of an intraclass correlation coefficient revealed that 1.8% of the variance in mutualism and 5.5% of the variance in domination exist between states. Following the formula offered by Snijders and Bosker (1999), the percentage of variance explained by modernization at both levels of the model (i.e., within states and between states) was calculated. Within states, the sociodemographic indicators had a negligible impact. However, at the macro level, these variables account for between 43% and 77% of the variance in mean value orientation scoring across states, revealing substantial influences.

These results are illustrated in Fig. 8.1, which displays state-level relationships between percent above median income and percent of mutualists within states. Notice that graphically this relationships shows a strong linear trend. Higher proportions of mutualists (and lower proportions of those who score

Table 8.2 Results of multilevel modeling procedures to assess the effects of modernization variables on wildlife value orientations

	Within-state effect		Contextual effect		
	Estimate[a]	SE[b]	Estimate	SE	PVE[c]
Education → Mutualism	–	–	–	–	0.05/0.68
Education[d]	–0.05	0.01*	0.92	0.14*	–
Gender[e]	0.48	0.04*	–	–	–
Age[f]	0.00	0.00	–	–	–
Income → Mutualism	–	–	–	–	0.06/0.50
Income[g]	–0.06	0.01*	0.46	0.09*	–
Gender	0.47	0.04*	–	–	–
Age	0.00	0.00	–	–	–
Urbanization → Mutualism	–	–	–	–	0.05/0.43
Urbanization[h]	0.03	0.01*	0.11	0.03*	–
Gender	0.47	0.04*	–	–	–
Age	0.00	0.00	–	–	–
Education → Domination	–	–	–	–	0.13/0.77
Education	–0.08	0.01*	–1.49	0.23*	–
Gender	–0.61	0.04*	–	–	–
Age	0.01	0.00*	–	–	–
Income → Domination	–	–	–	–	0.12/0.68
Income	0.02	0.01*	–0.77	0.11*	–
Gender	–0.61	0.05*	–	–	–
Age	0.01	0.00*	–	–	–
Urbanization → Domination	–	–	–	–	0.14/0.65
Urbanization	–0.07	0.01*	–0.20	0.04*	–
Gender	–0.60	0.04*	–	–	–
Age	0.01	0.00*	–	–	–

[a] Estimates represent unstandardized regression coefficients computed in Mplus 4.2 (Muthén & Muthén 1998–2006).

[b] SE = Standard error associated with estimates (regression coefficients).

[c] PVE = proportion of variance explained. The first number signifies PVE at level 1 (individual wildlife value orientation scoring within states), while the second number represents PVE at level 2 (mean wildlife value orientation scoring across states).

[d] Respondents indicated which of the following represented the highest level of education they had achieved: "less than high school diploma," "high school diploma or equivalent," "2-year associates degree or trade school," "4-year college degree," "advanced degree beyond 4-year college degree."

[e] Response options: "male" (1), "female" (2).

[f] Recorded by the respondent in number of years.

[g] Respondents indicated which of the following represented their annual household income before taxes: "less than $10,000," "$10,000–$29,999," "$30,000–$49,999," "$50,000–$69,999," "$70,000–$89,999," "90,000–$109,999," "$110,000–$129,999," "$130,000–$149,999," "$150,000 or more."

[h] Respondents indicated which of the following described their current residence or community: "a farm or rural area," "small town/village with less than 5,000 people," "town with 5,000–9,999 people," "town with 10,000–24,999 people," "small city with 25,000–49,999 people," "city with 50,000–99,999 people," "city with 100,000–249,999 people," "large city with 250,000 or more people."

* $p < .05$.

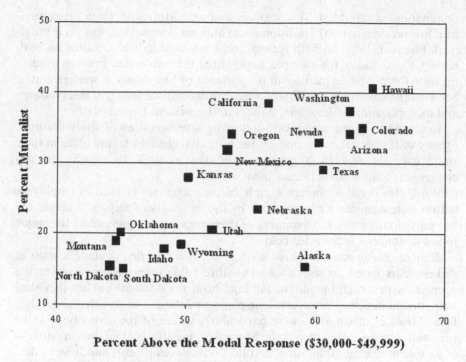

Fig. 8.1 Percent Mutualist by Income

high on the domination scale) in a state are associated with higher percentages of people above the median income category. Similar patterns would be reveled between percent mutualists and percentages residing in urban areas, and higher percentages of people who have attained greater than a high school education.

Wildlife Professionals Identify the Implications of Value Shift for Conservation

An important goal of the research reported here was to provide information to assist the wildlife management community in preparing for the future and, in particular, finding ways to better meet the demands of a changing public. Given that wildlife value orientations are changing in North America, what does that mean for wildlife managers? We thought that question is best answered by practicing wildlife professionals, so we chose, at the conclusion of this investigation, to seek their input through "futuring" workshops conducted in five of the participating states: Arizona, Idaho, Kansas, Oklahoma, and Wyoming. Workshops involved a two-stage process, beginning with a presentation of study results. To facilitate information transfer, findings were presented in the form of a stakeholder typology (instead of talking about mutualism and domination as variables). The typology included Mutualists (high on mutualism, low on domination) Traditionalists (high on domination, low on mutualism), Pluralists (high on both scales), and Distanced individuals (low on both scales). Conclusions, for example, highlighted the connection between modernization forces and an increase in proportions of Mutualists in western states. See Teel and Manfredo (2007b) for a more thorough discussion of this typology and its application in describing publics in the western United States.

In stage two of the workshops, following a presentation of study findings, agency staff were asked to consider the potential changes taking place in their states over the next 15–20 years and, in that context, to identify the top challenges facing wildlife management.

While this is not a random sample of state agencies or even of employees within state agencies, we were struck by the similarity of responses across the five participating states. A summary of these responses, organized by the major topics of concern, is provided below.

How can the agencies meet the needs of changing publics? A classification of stakeholders based on responses to wildlife value orientations provided the agencies with a useful guide for talking about the different publics they deal with, the changes they perceive taking place, and the day-to-day challenges they face. Those in urban areas were particularly aware of the diversity of value orientation types within their state. Moreover, the effects of migration patterns (e.g., due to energy exploration, second home development) and how to deal with newcomers to the state were discussed in this context. These changes were believed to be causing increased conflict (e.g., increased litigation) over wildlife issues, increased demand for non-traditional programs (e.g., wildlife viewing, non-game), and less support for traditional management approaches (e.g., hunting). These changes are also perceived as contributing to a growing lack of knowledge about and lack of interest in wildlife as people in urban areas are becoming more disconnected from the resource. Additional concerns centered around differences between agency culture and public values, the ability of the agencies to adapt to a changing clientele (e.g., need to be more proactive), and the underrepresentation of emerging interests, as well as minority populations, in agency decisions. While many indicated a need to embrace these underrepresented groups, some perceived this as a threat and were reluctant to consider change for fear of alienating traditional constituents. As an illustration, several workshop participants argued that the focus should be on trying to retain Traditionalists and Pluralists as opposed to trying to sway Mutualists into supporting the agency's efforts.

How can the agencies engage all publics in dealing with growing habitat loss? A major challenge discussed at these workshops was the growth and expansion of human populations, which results in a high rate of loss and fragmentation of wildlife habitat. Study findings promoted thinking about the diversity of human interests responsible for land use changes as well as the impact these changes could have on wildlife value orientations. As an example, a number of workshop participants argued that urbanization and land development have contributed to a loss of access to wildlife-related recreation opportunities and the ability for people to experience natural environments. This in turn raised concerns about growth of the Distanced group and resulted in identification of a need to provide opportunities for a broader segment of the public to enjoy wildlife (e.g., wildlife-related recreation opportunities in urban settings). In addition, agency staff advocated for more active engagement of *all* sectors of the public (i.e., all four wildlife value orientation types) as a necessary prerequisite to finding ways to deal with habitat loss. Engaging each of these sectors was viewed as critical in trying to balance the needs of wildlife with competing recreation interests and human uses (e.g., energy development, limited water resources) in the future. Additional issues cited in the context of habitat loss were urbanization and the "rural revitalization," whereby urban residents are migrating back to rural areas. These rural areas that once served as strongholds for the more traditional orientations toward the wildlife resource are changing and increasingly consist of a mix of potentially competing values.

How can the agencies deal with the rapid acceleration of human–wildlife conflict? Human population growth and expansion are contributing to a rise in human–wildlife conflict incidents. In particular, workshop discussions focused largely on an increase in human–wildlife conflict incidents in urban areas and places of expanding human settlement (e.g., the foothills or urban fringe) where a public uneducated about wildlife is brought in contact with them. Rapid urbanization and development of rural areas have resulted in increased human–wildlife interactions involving mutualists and other publics who may have less experience with, and knowledge about, the resource. Examples of incidents reported for these places included large carnivores such as mountain lions entering residential areas, moose in swimming pools, "nuisance" species such as geese, "pet" elk and deer, and wildlife injured in car accidents or as a result of being caught in fences or chased down by dogs. Due to changing publics affecting the level of support for traditional management approaches (e.g., lethal control) and a general lack of resources, agency staff indicated they feel increasingly limited in their ability to effectively address this growing problem of human–wildlife conflict. Their discussions built upon findings of this study that suggest a lack of support among Mutualists for management strategies resulting in death or harm to wildlife. Findings encouraged thinking about areas of consensus that might exist among wildlife value orientation types (e.g., greater consensus for using trained agency staff to remove "problem" animals) in determining appropriate strategies for addressing human–wildlife conflict.

How can the agencies combat declines in hunting? Trends in license sales discussed by the participating states indicated declines in hunting participation. For some agencies, this was evident in the form of a decline in actual numbers of licenses sold from one year to the next. For others, license sales may have remained somewhat stable in recent years, but the percent of the population that hunts has decreased. These declines raised concerns about funding, given the agencies' heavy reliance on hunting and fishing license sales as a source of revenue, and also about the potential loss of an important tradition and animal population management tool. A common question that arose among workshop participants was "How can we recruit more hunters?" However, in the context of this study's findings, which showed a strong connection between hunting and a more traditional wildlife value orientation, many participants recognized the difficulty or impossibility of reversing hunting-related trends in the future. Discussions instead centered on ways to better serve and retain the current population of hunters which, along with the public as a whole, is becoming increasingly diverse. Specifically, some argued that the agencies will need to find ways to appeal to the Pluralist or perhaps the Mutualist hunter in the future.

How can the agencies develop a secure source of funding for wildlife conservation? Related to the challenge of declining hunter numbers, western agencies are struggling with being able to secure and maintain a stable funding mechanism for wildlife conservation. By providing a better understanding of various publics, including their interests and how they may be reached in the future, this study encouraged thinking about ways to diversify agency funding structures. Workshop participants recognized a need to communicate more effectively with Mutualists and promote "more holistic" programs (e.g., that emphasize a diversity of species) that are likely to garner their support. Concerns focused on finding ways to get these individuals to pay for the wildlife they enjoy while at the same time ensuring the agencies do not alienate their traditional stakeholders who, up to now, have provided the bulk of monetary (and in many cases political) support for wildlife conservation. As one workshop participant stated, "50% of our constituency funds 95% of our activities. To broaden our services effectively, we must broaden our funding base."

Conclusion

The Changing Role of Wildlife Professionals

As acknowledged by the managers in our dialogue sessions, study findings suggest a changing future for wildlife management in North America. Society is becoming more focused on mutualism relationships with wildlife, and as a result the role of wildlife professionals is shifting. The traditional role of North American state wildlife agencies – defined, for example, by population-level management through provision of hunting and fishing opportunities – will be augmented by a role in which they enforce codes of conduct in human–wildlife relationships. In this emerging role, the duty of the professional is to deal with wildlife that "get out of line" or that are "bad" because they behave in a way that is outside human expectations of what is acceptable.

Stated another way, the agency will enforce sanctions against violators of human–wildlife interaction norms. It is quite analogous to the situation where police enforce legal norms of expected behavior in human interactions. To illustrate, a representative of the Utah Division of Wildlife Resources was recently describing the steps his agency took in the aftermath of an incident in which a bear attacked and killed a 13-year-old boy at a campground. To paraphrase his description, "Our procedures dictate we must treat the scene of a mauling just like the police treat the scene of a murder. We roped off the area and began to collect evidence. We sent out warning that there was a dangerous bear in the area and set out with dogs in an effort to track it down." In another example, a bear recently wandered into the town of Fort Collins, Colorado, ate berries from bushes in a residential area and climbed a large cottonwood tree to sleep during the day. The bear attracted a crowd, and the agency ultimately tranquilized and relocated it. Should the bear return, the agency announced, it would be destroyed per the agency's policy. These are just two of a growing number of instances where state wildlife agencies have been called upon to maintain desired human–wildlife relationships. There are, of course, "uneducated" or "bad" humans to deal with in these relationships, and they too are the target of agency efforts. This would include people who attempt to feed wildlife, who adopt orphaned wildlife, who create a situation around their homes that attracts predatory wildlife, who attempt to get close to wildlife while in a refuge like a national park. Increasingly, state wildlife officials will become the educators about norms of expected behavior and the "police" that enforce human–wildlife interactions.

Generalizations About Mutualism Beyond North America

Among most hunter-gatherer societies, people saw themselves as descended from the same ancestor as non-human animals in their clan. In post-modern society, many people perceive a close bond and reciprocity in relationships with some non-human animals. Although the forms of societal organization are quite different in this comparison, people's cognitions in both societies reveal elements of a mutualism value orientation toward animals. Is there any real similarity between these societies in their relationships with wildlife? Certainly, there would be little similarity in the specific beliefs about wildlife or the ways of treating wildlife, nor would human cognitions serve the same purposes. Those in hunter-gatherer societies are in close daily contact with natural environments, and their survival is linked to successful adaptation to these environments. Value orientations toward wildlife would therefore be tied, in some degree, to the functional purposes of environmental adaptation. In contrast, in post-modern society, people are increasingly removed from natural environments, and value orientations toward wildlife appear to play a small role in environmental adaptation.

The similarity between these situations, however, is in how humans find explanations for their surroundings. That is, people use what they know about their own social arrangements in finding predictability and understanding life with non-human animals (Willis, 1990). As noted by Mary Douglas (1990, p. 33), "...how could we think about how animals relate to one another except on the basis of our own relationships?" The form of desired relationships with animals is related to the nature of human relationships. In post-modern life, there is a greater emphasis on belongingness needs, self-expressive values (Inglehart, 1997), and egalitarian relationships, and day-to-day life is depicted as a "game against people" (Bell, 1973). We believe these conditions that have redefined human relationships (and their causes through modernization) have also given rise to mutualism as a new basis for human–wildlife interactions in post-modern society.

Is the rise of mutualism and its association with modernization a worldwide phenomenon? That would be a reasonable hypothesis given findings by both Inglehart (1997) and Schwartz (2006) who showed an association between level of modernization and the composition of values within a society. A team of researchers took the first step toward examining the prevalence of mutualism in societies outside North America by using qualitative techniques in a cross-cultural study of wildlife value orientations (Dayer, Stinchfield, & Manfredo, 2007; Teel, Manfredo, & Stinchfield, 2007). The principal objective of the study was to develop and test a technique that could be used in making cross-cultural comparisons. Findings showed interesting contrasts among the countries examined. Mutualism was quite prevalent in the Netherlands (Jacobs, 2007), Thailand (Tanakanjana & Saranet, 2007), and Estonia (Raadik & Cottrell, 2007). It was also present in the Chinese sample, but was expressed less frequently than materialism concerns (Zinn & Shen, 2007). Kaczensky (2007) contends that the mutualism orientation detected in Mongolia through this investigation is not the product of recent changes, but rather is part of a long-standing ideology. She suggested that a shift toward mutualism would not be a simple linear trend driven by socioeconomic development. Her conclusion is consistent with Inglehart and Baker (2000) who proposed that the global shift toward self-expressive values is nested within the religious tradition of countries. A refinement of the mutualism concept, to accurately reflect its expression in hunter-gatherer, pastoral, and more developed societies, might be helpful in that regard. Further explorations of the rise of mutualism will help us understand and manage human–wildlife relationships.

Summary

- HDW research will improve its utility by improving its conceptual foundation. Advancements can be made in commonly used concepts in HDW research (attitudes, values, norms), areas that have previously been neglected (emotions), and in the integration across social science approaches.

- In an illustration of new directions, a macro–micro level study is presented that proposes the rise of a mutualism value orientation toward wildlife in North America.
- In the micro model, ideology is introduced to the value–attitude–behavior hierarchy. Ideology is proposed to orient values, giving specific meaning to abstract ideals. Egalitarian ideology gives rise to a mutualism wildlife value orientation, and a domination ideology gives rise to a domination wildlife value orientation.
- The macro model proposes that modernization brings about new lifestyles that change the motivational basis for engagement with wildlife. As a result, with modernization, the domination wildlife value orientation is giving way to a mutualism view of the wildlife resource.
- Data to test these ideas come from a study conducted in 19 of the western United States. Findings show (a) strong prediction of attitudes toward wildlife-related issues from wildlife value orientations; (b) support for relationships specified in the full model values →wildlife value orientations → attitudes/ behaviors toward wildlife; (c) a strong macro-level effect (while controlling for individual-level relationships) from variables indicative of modernization (income, urbanization, education) on wildlife value orientations. Findings are consistent with the macro–micro model proposed in the study.
- With increasing emphasis on mutualism as the basis for relationships with wildlife, the role of state-level wildlife agencies will shift from population management through hunting and angling, to enforcement of the norms of human–wildlife relationships.
- It is reasonable to propose the global rise of a mutualism wildlife value orientation through forces of modernization, though preliminary cross-cultural investigations suggest the emergence of this orientation would take shape within the specific cultural traditions of a country.

Management Implications

The implications of the study described in this chapter were explored through workshops conducted with wildlife professionals in five different states. Results raised five basic questions:

1. How can agencies meet the needs of a changing public?
2. How can agencies engage all publics in dealing with growing habitat loss?
3. How can agencies deal with the rapid acceleration of human–wildlife conflict, especially as it relates to people seeking a mutualism relationship with wildlife?
4. How can agencies combat the associated declines in recreational hunting?
5. As traditional sources of funding dwindle due to losses of license sale revenue, how can agencies develop a secure source of funding for wildlife management activities?

Study results provided a framework for thinking about how to respond to these and other challenges in preparing for the future of wildlife management in North America.

References

Baron, R. M., & Kenny, D. A. (1986). The moderator-mediator variable distinction in social psychological research: Conceptual, strategic, and statistical considerations. *Journal of Personality and Social Psychology, 51*, 1173–1182.

Baumeister, R. F., & Leary, M. R. (1995). The need to belong: Desire for interpersonal attachments as a fundamental human motivation. *Psychological Bulletin, 117*, 497–529.

Bell, D. (1973). *The coming of post-industrial society: A venture in social forecasting*. New York: Basic Books.

Bright, A., Manfredo, M. J., & Fulton, D. (2000). Segmenting the public: An application of value orientations to wildlife planning in Colorado. *Wildlife Society Bulletin, 28*(1), 218–226.

Brown, T. L., Connelly, N. A., & Decker, D. J. (2006). *Participation in and orientation of wildlife professionals toward consumptive wildlife use: A resurvey*. HDRU Series Report No. 06-1. Ithaca, NY: Human Dimensions Research Unit.

Buttel, F. H., & Humphrey, C. R. (2002). Sociological theory and the natural environment. In R. E. Dunlap, & W. Michelson (Eds.), *Handbook of environmental sociology* (pp. 33–69). Westport, CT: Greenwood Press.

Catton, W. R. Jr., & Dunlap, R. (1980). A new ecological paradigm for post-exuberant sociology. *American Behavioral Scientist, 24*, 15–47.

Cochrane, W. W. (1993). *The Development of American Agriculture: A Historical Analysis*. Minneapolis: University of Minnesota Press.

Dayer, A. A., Stinchfield, H. M., & Manfredo, M. J. (2007). Stories about wildlife: Developing an instrument for identifying wildlife value orientations cross-culturally. *Human Dimensions of Wildlife, 12*(5), 307–315.

Dillman, D.A., & Tremblay, K. R. (1977). The quality of life in rural America. *Annals of the American Academy of Political and Social Science, 429*, 115–129.

Dizard, J. E. (2003). *Mortal stakes: Hunters and hunting in contemporary America*. Amherst, MA: University of Massachusetts Press.

Douglas, M. (1990). The pangolin revisited: A new approach to animal symbolism. In R. G. Willis (Ed.), *Signifying Animals: Human Meaning in the Natural World* (pp. 25–36). London: Unwin Hyman.

Dunlap, R. E. (2002). An enduring concern: Light stays green for environmental protection. *Policy Perspective, September/October, 13*(5), 10–14.

Fiske, A. P. (1992). The four elementary forms of sociality: Framework for a unified theory of social relations. *Psychological Review, 99*, 689–723.

Fulton, D. C., Manfredo, M. J., & Lipscomb, J. (1996). Wildlife value orientations: A conceptual and measurement approach. *Human Dimensions of Wildlife, 1*(2), 24–47.

Hand, C. M., & Van Liere, K. D. (1984). Religion, mastery-over-nature, and environmental concern. *Social Forces, 63*, 555–570.

Heberlein, T. A. (1991). Changing attitudes and funding for wildlife: Preserving the sporthunter. *Wildlife Society Bulletin, 19*, 528–534.

Herman, D. J. (2003). The hunter's aim: The cultural politics of American sport hunters, 1880–1910. *Journal of Leisure Research, 35*, 455–474.

Hofstede, G. (2002). The pitfalls of cross-national survey research: A reply to the article by Spector et al. on the psychometric properties of the Hofstede Values Survey Module 1994. *Applied Psychology: An International Review, 51*, 170–173.

Homer, P. M., & Kahle, L. R. (1988). A structural equation test of the Value-Attitude-Behavior Hierarchy. *Journal of Personality and Social Psychology, 54*(4), 638–646.

Inglehart, R. (1990). *Culture shift in advanced industrial societies.* Princeton University Press.

Inglehart, R. (1997). *Modernization and postmodernization.* Princeton, New Jersey: Princeton University Press.

Inglehart, R., & Baker, W. E. (2000). Modernization, cultural change, and the persistence of traditional values. *American Sociological Review, 65*(1), 19–51.

Inglehart, R., & Welzel, C. (2005). *Modernization, cultural change and Democracy: The human development sequence.* New York: Cambridge University Press.

Ingold, T. (1994). From trust to domination: An alternative history of human-animal relations. In A. Manning, & J. Serpell (Eds.), *Animals and human society: Changing perspectives* (pp. 1–22). New York: Routledge.

Jacobs, M. H. (2007). Wildlife value orientations in the Netherlands. *Human Dimensions of Wildlife, 12*(5), 359–365.

Jacobson, L. (2006). Ballot measure wrap-up. *The Rothenberg Political Report.* Retrieved Tuesday, November 21, 2006 from http://rothenbergpoliticalreport.blogspot.com/2006/11/ballot-measure-wrap-up.html.

Kaczensky, P. (2007). Wildlife value orientations of rural Mongolians. *Human Dimensions of Wildlife, 12*(5), 317–329.

Katcher, A., & Wilkins, G. (1993). Dialogue with animals: Its nature and culture. In S. R. Kellert, & E. O. Wilson (Eds.), *The Biophilia Hypothesis* (pp. 173–200). Washington, DC: Island Press.

Kellert, S. R. (1976). Perceptions of animals in American society. *Transactions of the North American Wildlife and Natural Resources Conference, 41*, 533–546.

Kluckhohn, C. (1951). Values and value orientations in the Theory of Action. In T. Parsons, & E. A. Shils (Eds.), *Toward a general Theory of Action* (pp. 388–433). Cambridge: Harvard University Press.

Kluckhohn, F. R., & Strodtbeck, F. (1961). *Variations in value orientations.* Evanston: Row, Peterson, & Co.

Maio, G. R., Olson, J. M., Bernard, M. M., & Luke, M. A. (2003). Ideologies, values, and behavior. In J. Delamater (Ed.), *Handbook of social psychology* (pp. 283–308). London: Springer.

Manfredo, M. J., & Zinn, H. C. (1996). Population change and its implications for wildlife management in the new west: A case study of Colorado. *Human Dimensions of Wildlife, 1*, 62–74.

Manfredo, M. J., Teel, T. L., & Bright, A. D. (2004). Application of the concepts of values and attitudes in human dimensions of natural resources research. In M. J. Manfredo, J. J. Vaske, D. Field, & P. J. Brown (Eds.), *Society and Natural Resources: A summary of knowledge prepared for the 10th International Symposium on Society and Natural Resources* (pp. 271–282). Jefferson, MO: Modern Litho.

Manfredo, M. J., Teel, T., & Henry, K. (2007). Modernization and the rise of mutualism in human-wildlife relationships. Unpublished manuscript.

Maslow, A. H. (1954). *Motivation and personality.* New York: Harper & Row.

Milton, K. (1996). *Environmentalism and cultural theory: Exploring the role of Anthropology in environmental discourse.* London: Routledge.

Minnis, D. L. (1998). Wildlife policy-making by the electorate: An overview of citizen-sponsored ballot measures on hunting and trapping. *Wildlife Society Bulletin, 26*, 75–83.

Mithen, S. (1996). *The prehistory of the mind.* London: Thames & Hudson.

Muth, R. M., & Jamison, W. V. (2000). On the destiny of deer camps and duck blinds: The rise of the animal rights movement and the future of wildlife conservation. *Wildlife Society Bulletin, 28*, 841–851.

Organ, J. F., & Fritzell, E. K. (2000). Trends in consumptive recreation and the wildlife profession. *Wildlife Society Bulletin, 28*, 780–787.

Peyton, R. B. (2000). Wildlife management: Cropping to manage or managing to crop? *Wildlife Society Bulletin, 28*, 774–779.

Pratto, F. (1999). The puzzle of continuing group inequality: Piecing together psychological, social and cultural forces in social dominance theory. *Advances in Experimental Social Psychology, 31*, 191–263.

Raadik, J., & Cottrell, S. (2007). Wildlife value orientations: An Estonian case study. *Human Dimensions of Wildlife, 12*(5), 347–357.

Raudenbush, S. W., & Bryk, A. S. (2002). *Hierarchical linear models: Applications and data analysis methods* (2nd ed.). Thousand Oaks, CA: Sage Publications.

Rokeach, M. (1973). *The nature of human values*. New York: Free Press.

Schwartz, S. H. (1994). Beyond individualism/collectivism: New cultural dimensions of values. In U. Kim, H. C. Triandis, C. Kagitcibasi, S. Choi, & G. Yoon (Eds.), *Individualism and collectivism: Theory, methods, and applications* (Vol. 18, Cross Cultural Research and Methodology Series, pp. 85–119). London: Sage.

Schwartz, S. H. (2004). Basic human values: Their content and structure across cultures. In A. Tamayo, & J. Porto (Eds.), Vialores *e Trabalho*. Brazil: Editora Universidade de Brasilis.

Schwartz, S. H. (2006). A theory of cultural value orientations: Explication and applications. *Comparative Sociology, 5*, 136–182.

Serpell, J. A. (2003). Anthropomorphism and anthropomorphic selection – Beyond the 'cute response'. *Society and Animals, 11*(1), 83–100.

Sidanius, J. (1993). The psychology of group conflict and the dynamics of oppression: A social dominance perspective. In S. Iyengar, & W. McGuire (Eds.), *Explorations in Political Psychology* (pp. 183–219). Durham, NC: Duke University Press.

Sidanius, J., Pratto, F., van Laar, C., & Levin, S. (2004). Social dominance theory: Its agenda and method. *Political Psychology, 25*, 845–880.

Snijders, T. A., & Bosker, R. J. (1999). *Multilevel analysis: An introduction to basic and advanced multilevel modeling*. Thousand Oaks, CA: Sage Publications.

Tanakanjana, N., & Saranet, S. (2007). Wildlife value orientations in Thailand: preliminary findings. *Human Dimensions of Wildlife, 12*(5), 339–345.

Teel, T. L., Dayer, A. A., Manfredo, M. J., & Bright, A. D. (2005). *Regional results from the research project entitled* "Wildlife Values in the West." (Project Rep. No. 58). Project Report for the Western Association of Fish and Wildlife Agencies. Fort Collins, CO: Colorado State University, Human Dimensions in Natural Resources Unit.

Teel, T. L., Manfredo, M. J., & Stinchfield, H. S. (2005). The need and theoretical basis for exploring wildlife value orientations cross-culturally. *Human Dimensions of Wildlife, 12*(5), 297–307.

Teel, T. L., & Manfredo, M. J. (2007a). Exploring the predictive validity of the wildlife value orientation concept across a diverse set of wildlife-related issues. Unpublished manuscript.

Teel, T. L. & Manfredo, M. J. (2007b). Understanding the diversity of public interests in wildlife conservation. Unpublished manuscript.

Triandis, H. C. (1995). *Individualism and collectivism*. Boulder, CO: Westview.

U.S. Fish and Wildlife Service. (2007). *2006 National survey of fishing, hunting, and wildlife-associated recreation: National overview*. Washington, DC: U.S. Fish and Wildlife Service.

Vaske, J. J., Shelby, L. B., & Manfredo, M. J. (2006). Bibliometric reflections on the first decade of *Human Dimensions of Wildlife*. *Human Dimensions of Wildlife, 11*(2), 79–87.

Vining, J. (2003). The connection to other animals and caring for nature. *Human Ecology Review, 10*(2), 87–99.

Werner, C. M., Brown, B. B., & Altman, I. (1997). Environmental psychology. In J. Berry, M. H. Segal, & C. Kagitibasi (Eds.), *Handbook of cross-cultural psychology* (Vol. 3, Social Behavior and Applications, pp. 255–290). Boston: Allan and Bacon.

White, L. (1967). The historical roots of our ecologic crisis. *Science, 155*, 1203–1207.

Wildavsky, A. B. (1991). *The rise of radical egalitarianism*. Washington, DC: The American University Press.

Willis, R. (Ed.). (1990). *Signifying animals: Human meaning in the natural world*. London: Routledge.

Zinn, H. C., & Shen, X. S. (2007). Wildlife value orientations in China. *Human Dimensions of Wildlife, 12*(5), 331–338.

Appendix: Item Scales for Developing Wildlife Value Orientations in North America

The approach to developing wildlife value orientation item scales was guided by standard principles of psychological testing (Murphy & Davidshofer, 2005). With this approach, scales are developed to attend to concerns of content validity, reliability, construct validity, and predictive validity. The discussion of item scales below is developed in line with each of these topics. In presenting this discussion, it is important to emphasize that reliability and validity *are not* characteristics of item scales such as those presented here (Thompson, 2003). Instead, they are characteristics of *a specific use of an instrument*. An important implication of this idea is that each use of an instrument such as ours requires reliability and validity assessments. No amount of prior testing absolves the researcher from such basic considerations.

Content Validity

The notion of content validity deals with whether the items in one's instrument adequately assess all the critical aspects of the concept they are intended to. Evaluating content validity is largely a subjective assessment. In developing the items reported here, we have relied on prior literature, individual interviews, and several empirical investigations in which items were tested and refined. (For descriptions of these formative studies, see Fulton, Manfredo, & Lipscomb, 1996; Fulton, Pate, & Manfredo, 1995; Manfredo, Teel, & Bright, 2003; Teel, Dayer, Manfredo, & Bright, 2005.) The purpose of scale development is to match theoretical constructs with specific measures. In this case, the items presented here are intended to assess beliefs about appropriate modes of conduct regarding wildlife and beliefs about an ideal world in human–wildlife relationships. Both of these belief sets are theorized to be critical components of the wildlife value orientation concept.

Tests of Reliability

Reliability deals with the extent to which the results of a test are consistent and repeatable. As an illustration, a measuring stick that yields the same result in five separate tries when assessing person A's height has produced reliable findings. This may be a relatively simple task when measuring physical characteristics; however, when taking measurements in psychology, where the characteristics of interest are not readily observable, reliability can be challenging.

Our development of item scales was guided by classical test theory of reliability. This approach is quite common and has been the foundation for reliability considerations in psychological measurement for over 80 years (Kline, 2005). This approach suggests that an individual's response on a survey item is composed of a *true score* and *error*. Error would include the effect of many things such as temporary conditions, situational factors, and response biases that are reflected in how a person answers questions. The ultimate goal of scaling is to develop items that obtain responses predominated by the enduring *true score effects* and in which error is minimized and random across many individuals. At the practical level, this theory leads to the notion that the measurement of a concept requires the use of multiple items that are all directed toward assessment of that same concept. That is, the best way to lower the impact of error effects is to use multiple items for measuring a single concept.

Another benefit of using multi-item scales is that reliability can actually be estimated when you have multiple measures of the same thing. If a group of items is developed to measure a specific conceptual topic, the extent to which consistent results are obtained across all items indicates the reliability of the scale's use. *Cronbach's alpha* is a statistic available to measure internal consistency of a group of items that builds upon this notion. Theoretically, Cronbach's alpha is defined as the correlation between the scale items used and the population of possible items. More practically, alpha is used to guide instrument development; items are retained to construct a scale if they produce results consistent with other items and contribute to improvement in the overall alpha score. Nunnally and Bernstein (1994) suggested that an alpha of around 0.70 be used as a cutoff in defining the reliability of a scale; i.e., an alpha of at least 0.70 means the scale has acceptable reliability. Table 1 shows the alpha reliability coefficients from our use of scales in the 2004 application of the wildlife value orientation instrument (Teel et al., 2005).

Construct Validity

Constructs are hypothetical abstractions that give predictability to real observable things or events. As noted by Murphy and Davidshofer (2005), constructs are the essence of science. Construct development in psychology involves (a) the

Table 1 Reliability results for value orientations and their wildlife basic beliefs

Value orientation	Basic belief	Cronbach's alpha
Domination	–	0.83
–	Appropriate use	0.78
–	Hunting belief	0.80
Mutualism		0.86
–	Social affiliation belief	0.82
–	Caring belief	0.80

To see the items that measure each basic belief within each value orientation, see the second column of Table 2.

description of a concept in ways that make it distinct from other concepts and (b) the identification of the behaviors that are related to the concept.

Questions regarding the *construct validity* of an instrument, such as the one presented here, ask whether the items provide a good measure of the proposed concept. The challenge of examining construct validity is that we do not have a real, true measurement to which we can compare our test scores. Hence, construct validation is largely an indirect, inferential exercise. For example, if three separate means of measuring a construct obtain very similar results, we have indirect evidence in favor of construct validity.

While there are numerous, highly sophisticated statistical techniques for conducting this test, the question being addressed here is basic. Do items within a group intended to measure a construct (e.g., mutualism) have strong correlations with one another? Further, do these items have low correlations with items in groups that are intended to measure different constructs? For example, do items intended to measure mutualism correlate more strongly with one another than they do with items in the domination scale? Our investigation did, in fact, find that the average correlation of items *within* wildlife value orientation scales was significantly stronger than the correlation with items *outside* the wildlife value orientations (Teel et al., 2005).

More sophisticated testing of these items leads to similar conclusions. Confirmatory factor analysis (CFA) shows whether our hypothesized item groupings provided a good fit that explains the scoring variance on our instrument. Results taken from Teel et al. (2005) showed standardized factor loadings ranging from 0.42 to 0.85 (all t values were significant at $p < 0.001$) for items comprising basic belief dimensions (Table 2) and from 0.53 to 0.86 (all t values were significant at $p < 0.001$) for loadings of belief dimension scales on their respective domination and mutualism value orientations (Table 2). These findings support the idea that our item groupings reflect distinct scoring patterns consistent with our theoretical framework. More specifically, people responded similarly across items intended to measure a domination orientation, and this response was different from individual response patterns on the mutualism items.

Predictive validity, a distinct type of construct validity, addresses the question of whether proposed concepts predict the outcomes they are theorized to

Table 2 CFA results for wildlife value orientations

Basic Belief	Value Orientation and the Items Measuring it[1]	Factor Loading[2]	SE[3]	t[4]
	Domination			
Appropriate use	–	0.86	0.02	75.79
–	• Humans should manage fish and wildlife populations so that humans benefit.	0.57	0.02	60.73
	• The needs of humans should take priority over fish and wildlife protection.	0.65	0.02	70.85
	• It is acceptable for people to kill wildlife if they think it poses a threat to their life.	0.54	0.01	56.57
	• It is acceptable for people to kill wildlife if they think it poses a threat to their property.	0.68	0.02	75.74
	• It is acceptable to use fish and wildlife in research even if it may harm or kill some animals.	0.54	0.02	56.52
	• Fish and wildlife are on earth primarily for people to use.	0.67	0.02	73.23
	–	–	–	–
Hunting belief	–	0.53	0.01	53.65
–	• We should strive for a world where there's an abundance of fish and wildlife for hunting and fishing.	0.51	0.02	53.66
	• Hunting is cruel and inhumane to the animals.[R]	0.79	0.02	93.22
	• Hunting does not respect the lives of animals.[R]	0.80	0.02	94.71
	• People who want to hunt should be provided the opportunity to do so.	0.74	0.01	85.06
	Mutualism	–	–	–
Social affiliation		0.82	0.02	83.61
–	• We should strive for a world where humans and fish and wildlife can live side by side without fear.	0.57	0.02	62.10
	• I view all living things as part of one big family.	0.73	0.02	85.15
	• Animals should have rights similar to the rights of humans.	0.81	0.02	99.22
	• Wildlife are like my family and I want to protect them.	0.82	0.02	100.80
	–	–	–	–

Table 2 (continued)

Basic Belief	Value Orientation and the Items Measuring it[1]	Factor Loading[2]	SE[3]	t[4]
Caring beliefs	–	0.67	0.01	69.77
–	• I care about animals as much as I do other people.	0.53	0.02	57.59
	• It would be more rewarding to me to help animals rather than people.	0.42	0.02	43.55
	• I take great comfort in the relationships I have with animals.	0.84	0.01	103.41
	• I feel a strong emotional bond with animals.	0.72	0.02	83.76
	• I value the sense of companionship I receive from animals.	0.85	0.01	105.61

[1]Item response scales ranged from 1 = *strongly disagree* to 7 = *strongly agree*.
[2]Numbers represent standardized factor loadings calculated in the CFA. Following the initial CFA which calculated loadings for items on their respective belief dimensions, we created belief dimension scales (mean composites of individual items) to examine how the latter loaded on the hypothesized value orientations.
[3]SE = Standard error associated with factor loadings.
[4]All *t*-values calculated in the CFA were significant at $p < 0.001$.
[R]Item was reverse-coded prior to analysis.

predict. As evidence of predictive validity, study results reported in Chapter 8 show strong associations in the direction predicted for the relationship between wildlife value orientations and wildlife-related attitudinal and behavioral outcomes.

References

Fulton, D. C., Pate, J., & Manfredo, M. J. (1995). Colorado residents' attitudes toward trapping in Colorado (Project Rep. No. 23, Project Rep. for the Colorado Division of Wildlife). Fort Collins: Colorado State University, Human Dimensions in Natural Resources Unit.

Fulton, D. C., Manfredo, M. J., & Lipscomb, J. (1996). Wildlife value orientations: A conceptual and measurement approach. *Human Dimensions of Wildlife, 1*(2), 24–47.

Kline, T. J. B. (2005). *Psychological testing: A practical approach to design and evaluation.* Thousand Oaks, CA: Sage Publications.

Murphy, K. R., and Davidshofer, C. O. (2005). *Psychological testing: principles and applications* (6th ed.). Upper Saddle River, New Jersey: Pearson Prentice Hall.

Manfredo, M. J., Teel, T. L., and Bright, A. D. (2003). Why are public values toward wildlife changing? *Human Dimensions of Wildlife, 8*, 287–306.

Nunnally, J. C., and Bernstein, I. H. (1994). *Psychometric theory* (3rd ed.). New York, NY: McGraw Hill.

Teel, T. L., Dayer, A. A., Manfredo, M. J., and Bright, A. D. (2005). Regional results from the research project entitled 'Wildlife values in the west.' (Project Re. No. 58, Project Rep. for the Western Association of Fish and Wildlife Agencies). Fort Collins, CO: Colorado State University, Human Dimensions in Natural Resources Unit.

Thompson, B. (2003). Understanding reliability and coefficient alpha, really. In B. Thompson (Ed.), *Score Reliability: Contemporary thinking on reliability issues* (pp. 3–23). Thousand Oaks, CA: Sage Publications.

Index